Springer Theses

Recognizing Outstanding Ph.D. Research

For further volumes:
http://www.springer.com/series/8790

Aims and Scope

The series "Springer Theses" brings together a selection of the very best Ph.D. theses from around the world and across the physical sciences. Nominated and endorsed by two recognized specialists, each published volume has been selected for its scientific excellence and the high impact of its contents for the pertinent field of research. For greater accessibility to non-specialists, the published versions include an extended introduction, as well as a foreword by the student's supervisor explaining the special relevance of the work for the field. As a whole, the series will provide a valuable resource both for newcomers to the research fields described, and for other scientists seeking detailed background information on special questions. Finally, it provides an accredited documentation of the valuable contributions made by today's younger generation of scientists.

Theses are accepted into the series by invited nomination only and must fulfill all of the following criteria

- They must be written in good English.
- The topic of should fall within the confines of Chemistry, Physics and related interdisciplinary fields such as Materials, Nanoscience, Chemical Engineering, Complex Systems and Biophysics.
- The work reported in the thesis must represent a significant scientific advance.
- If the thesis includes previously published material, permission to reproduce this must be gained from the respective copyright holder.
- They must have been examined and passed during the 12 months prior to nomination.
- Each thesis should include a foreword by the supervisor outlining the significance of its content.
- The theses should have a clearly defined structure including an introduction accessible to scientists not expert in that particular field.

Mateusz Wielopolski

Testing Molecular Wires

A Photophysical and Quantum Chemical Assay

Doctoral Thesis accepted by
Friedrich-Alexander
University of Erlangen-Nuernberg, Germany

Author
Dr. Mateusz Wielopolski
Department of Chemistry and Pharmacy
Computer Chemistry Center (CCC)
University of Erlangen
Egerlandstr. 3
91058 Erlangen, Germany
e-mail: mateusz.wielopolski@chemie.
uni-erlangen.de

Supervisor
Prof. Dr. Dirk M. Guldi
Department of Chemistry and Pharmacy
University of Erlangen
Egerlandstr. 3
91058 Erlangen, Germany
e-mail: guldi@chemie.uni-erlangen.de

ISSN 2190-5053

e-ISSN 2190-5061

ISBN 978-3-642-14739-5

e-ISBN 978-3-642-14740-1

DOI 10.1007/978-3-642-14740-1

Springer Heidelberg Dordrecht London New York

Library of Congress Control Number: 2010929574

© Springer-Verlag Berlin Heidelberg 2010

This work is subject to copyright. All rights are reserved, whether the whole or part of the material is concerned, specifically the rights of translation, reprinting, reuse of illustrations, recitation, broadcasting, reproduction on microfilm or in any other way, and storage in data banks. Duplication of this publication or parts thereof is permitted only under the provisions of the German Copyright Law of September 9, 1965, in its current version, and permission for use must always be obtained from Springer. Violations are liable to prosecution under the German Copyright Law.

The use of general descriptive names, registered names, trademarks, etc. in this publication does not imply, even in the absence of a specific statement, that such names are exempt from the relevant protective laws and regulations and therefore free for general use.

Cover design: eStudio Calamar, Berlin/Figueres

Printed on acid-free paper

Springer is part of Springer Science+Business Media (www.springer.com)

For Maximus

Supervisor's Foreword

One of the major challenges in current chemistry is to find molecules able to move charges rapidly and efficiently from, for example, one terminus to another one under the control of an external electrical, electrochemical or photochemical stimulus. Nature has provided impressive examples of how these goals are achieved. The photosynthetic reaction center protein, for instance, rapidly moves electrons with near unity quantum efficiency across a lipid bilayer membrane using several redox cofactors, and thus, serves as a model for developing biomimetic analogues for applications in fields such as photovoltaic devices, molecular electronics and photonic materials. In this context, π-conjugated oligomeric molecular assemblies are of particular interest because they provide efficient electronic couplings between electroactive units - donor and acceptor termini - and display wire-like behavior. In order to make a molecule able to behave as an ideal molecular wire different requirements need to be fulfilled: i) matching between the donor (acceptor) and bridge energy levels, ii) a good electronic coupling between the electron donor and acceptor units via the bridge orbitals, and iii) a small attenuation factor.

Among the many different π-conjugated oligomers, *oligo*(p-phenylenevinylenes) (*o*PPV), have emerged as a particularly promising model system that helps to comprehend/rationalize the basic features of polymeric poly(p-phenylenevinylenes) and also as a versatile building block for novel materials with chemically tailored properties. In this context, intramolecular electron transfer along conjugated chains of *o*PPV has been tested in several donor-bridge-acceptor conjugates involving anilines, porphyrins, and ferrocenes, as electron donors, on one side, and fullerenes, on the other side, as the electron acceptor. In fact, in a pioneering study tetracene—as electron donor—and pyromellitimide—as electron acceptor—were connected via *o*PPV of increasing length. This work has demonstrated the importance of energy matching between the donor and bridge components for achieving a molecular-wire behavior. Quantum-chemical calculations showed a competition between a direct superexchange process and a two-step "bridge-mediated" process, whose efficiency depends primarily on the length and nature of the conjugated bridge.

The work accomplished by Mateusz Wielopolski is a great asset to the aforementioned, namely the field of charge transport through organic π-conjugated molecules. It was embedded in the *Collaborative Research Center SFB 583, "Redox-Active Metal Complexes: Control of Reactivity via Molecular Architecture"* at the University Erlangen-Nuremberg, which provided a perfect environment through strong cooperation partners in physics and chemistry, experimentally as well as theoretically. Especially in the steadily growing field of molecular electronics and organic photovoltaics, the demand for potential alternatives to convential materials for the construction of electronic devices and organic solar cells underlines the significance of the quest after suitable materials to transport electricity without losses. It has been shown that the investigated molecular structures are suitable for distance-independent charge transport. More precisely, π-conjugated oligomeric systems were tested with regard to their ability to provide an efficient electronic coupling between the electroactive and display wire-like behavior. The different factors that qualify molecules as ideal molecular wires, such as matching between the donor (acceptor) and bridge energy levels, electronic couplings and small attenuation factors, were fine-tuned. Furthermore, the number of publications emerged from this thesis - vide infra - is outstanding and confirms the momentousness of this topic.

Erlangen, September 2010   Dirk M. Guldi

Acknowledgments

In the first place, I would like to dedicate a great thank you to **Prof. Dr. Dirk M. Guldi** and **Prof. Dr. Timothy Clark** for giving me the opportunity to accomplish the present PhD thesis as an interdisciplinary project between the two institutes of Physical Chemistry I and the Computer Chemie Centrum. With their continual interest and assistance from both ends—the experimental and theoretical—I was able to gather a very detailed scientific background for the investigated processes and their various characterization methods. By far more important is the fact that, whenever we had the chance to leave the scientific fields, our conversations turned out to be even more motivating and fruitful.

Additionally, I would like to thank **Prof. Dr. Andreas Hirsch** for taking the chair during the exam and his commitment to the musical side of science.

The synthetic groups in Madrid and in Durham deserve great thanks for their efforts in very reliable chemical synthesis of the investigated compounds. In this regard, it should be stressed that without the grandiose work of **Prof. Dr. Nazario Martin**, **Prof. Dr. Martin Bryce**, **Dr. Cornelia van der Pol** and **Dr. Salvatore Filippone**, none of the published and herein presented results would even be imaginable.

Great thanks to **Dr. Carmen Atienza-Castellanos** for the perfect instructions regarding various measurement techniques and scientific problems. Her preliminary works provided a solid basis for the obtained results in this thesis. Her steady support—even from the far Madrid—renders her not only a reliable colleague but also a good friend. Thanks also to **Dr. Harald Lanig** and **Dr. Nico van Eikema Hommes** for their support regarding any molecular modeling issues in praxis and theory, especially hard—and software problems. Great thanks are dedicated to **Dr. Guido Sauer**, **Dr. Axel Kahnt**, **Dr. Christian Ehli** and **Dr. Georg Brehm** for providing me a perfect working environment regarding femto- and nanosecond spectroscopy.

Another thank you is addressed to **Anna Wielopolski** for dismissing me from extended fatherhood duties and her support in everyday life. In the same way, thanks to my son, **Maximus**, since I appreciate his interest in fullerene-like architectures on various playgrounds. Moreover, I want to say thank you to my

Parents and **Family** for their steady support in any circumstances of life. Similarly, I would like to set **Fabian Spänig** apart who turned out to be the perfect room mate. Not only his immense knowledge, but also his social skills made him a very special person and friend to me.

Thank you, **Freddy** and **Kramer**, for the good times at university and, in particular, outside.

All my colleagues from the Guldi group deserve a huge THANK YOU for so much fun and such an enjoyable working atmosphere. In that sense, thank you **Bruno**, **Fabian W**, **Daniel**, **Droste**, **Vito**, **Shankara**, **Wolfgang**, **Christina**, **Anita**, **Jürgen**, **Gustavo**, **Silke**, **Esther**, **Andres**, **Christian Ö**, **Daniela**, **Renata**, **Sebastian**, **Anna**, **Gerd** and **Dirk**.

Equally important, the group members of Professor Clark earn great thanks for all the fun and fruitful discussions. Thus, thank you **Christof**, **Florian** and **Volker** (for the necessary motivation to do it fast), **Angela**, **Matthias Schwofi**, **Rene Weller**, **Tatyana**, **Hakan**, **Sebastian**, **Frank and Jr-Hung**.

Erlangen   Mateusz Wielopolski

Contents

Part I Introduction and Motivation

1 Introduction to Molecular Electronics 3
   1.1 Present Technology 6
   1.2 Limitations of Present Technology 6
   References 8

2 Motivation—Focusing on Molecular Wires 9
   References 10

Part II Theoretical Concepts

3 Concepts of Photoinduced Electron and Energy Transfer Processes Across Molecular Bridges 13
   3.1 Introduction 13
   3.2 Electron Transfer Mechanisms 14
      3.2.1 Superexchange 16
      3.2.2 Charge Hopping 19
      3.2.3 Interplay of Mechanisms 21
   3.3 Electronic Energy Transfer 21
      3.3.1 Coulombic Energy Transfer 23
      3.3.2 Exchange Energy Transfer 23
   References 24

4 Molecule-Assisted Transport of Charges and Energy Across Donor–Wire–Acceptor Junctions 27
   4.1 Mechanisms of Charge Transfer through Molecular Wires 28
      4.1.1 Superexchange Charge Transfer in Molecular Wires ... 29
      4.1.2 Sequential Charge Transfer in Molecular Wires 30

	4.2	Factors that Determine the Charge Transfer Mechanism	31
		4.2.1 Electronic Coupling	31
		4.2.2 Energy Matching	32
	4.3	Specific Aspects of Photoinduced Electron Transfer in Organic π-Conjugated Systems	34
		4.3.1 Background	34
		4.3.2 The Classical Marcus Theory	35
		4.3.3 Photoexcitation and Relaxation Processes in Solution	38
		4.3.4 Influence of the Solvation on the Electronic Relaxation Dynamics	48
	References		51
5	**Examples of Molecular Wire Systems**		55
	5.1	Oligo(phenylenevinylene)s	55
	5.2	Oligophenylenes	57
	5.3	Oligo(thiophene)s	58
	5.4	Photonic Wires	59
	References		60

Part III Results and Discussion

6	**Objective**		65
	References		70
7	**Instruments and Methods**		71
	7.1	Photophysics	71
		7.1.1 Absorption Spectroscopy	71
		7.1.2 Steady-state Emission	71
		7.1.3 Time-resolved Emission	71
		7.1.4 Femtosecond Transient Absorption Spectroscopy	72
		7.1.5 Nanosecond Laser Flash Photolysis	73
	7.2	Chemicals	73
	7.3	Molecular Modeling	74
	References		75
8	**Energy Transfer Systems**		77
	8.1	Linking two C_{60} Electron Acceptors to a Molecular Wire	77
		8.1.1 C_{60}–oPPE–C_{60}—A Representative Example for Efficient Energy Transfer	77
		8.1.2 Energy Transfer in C_{60}–$oligo$(fluorene)–C_{60}	81
	8.2	Tunable Excited State Deactivation	90
		8.2.1 Photophysics	92
	References		98

9	**Electron Transfer Systems**		99
	9.1 *p*-Phenyleneethynylene Molecular Wires		99
		9.1.1 *ex*TTF–*o*PPE–C_{60} Donor–Acceptor Conjugates	100
		9.1.2 H_2P/ZnP–*o*PPE–C_{60} Donor–Acceptor Conjugates	116
		9.1.3 Meta-Connectivity—Influence of Structure on Molecular Wire Properties	131
	9.2 *oligo*-Fluorene Molecular Wires		145
		9.2.1 *ex*TTF–*o*FL–C_{60} Donor–Acceptor Conjugates	146
		9.2.2 ZnP–*o*FL–C_{60} and Ferrocene–*o*FL–C_{60} Donor–Acceptor Conjugates	157
	References		171
10	**Conclusions and Outlook**		173
Curriculum Vitae			179
Publications			181

Part I
Introduction and Motivation

Chapter 1
Introduction to Molecular Electronics

In general, molecular electronics involves single or small groups of molecules in device-based fabrication for electronic components, such as wires, switches, memory and gain elements [1–4]. To this end, molecular electronics emerged as an area of research, stimulating the creative minds of scientists in a way that only few research topics have ever done in the past [5]. Simply the fact, that the *Science* magazine labeled the integration of molecules into functional electric circuits one of the revolutions of 2001 [6], motivates teams of chemists, engineers, materials scientists, physicists and computer experts initiate cooperations to ultimately convert this interdisciplinary field into commercially available products.

In view of current state of the art, namely, silicon-based technology, molecular electronics definitely exceed the expectations of a single product line. Going back, for instance, to 1960 silicon-based electronics were nearly exclusively considered as a simple replacement for the vacuum tube. However, it would have been myopic to limit the potential of silicon in that field of research. In fact, silicon constituted a technological platform, which evolved into the development of various products, most of them unfathomable at the time. Similarly, molecular electronics may be considered as a platform technology, rather than a single product line, which may give rise to many industrial products which are currently unforeseeable.

However, arguments on a mesoscopic scale become huge. Considering that the integration of molecular materials into electronic devices [7] requires films or crystals, which contain many trillions of molecules per functional unit, their properties should be explored on the macroscopic scale. On the other hand, a clear distinction has to be made regarding molecular scale electronics with one to a few thousand molecules per device. To visualize this fact, thin film transistors (TFTs) and polymer-based light emitting diodes (LEDs) are also composed of molecular materials for electronics. The mean grain size at which they should exhibit their function is approximately in the $2\,\mu m$ region. Such size domains are already in reach even by silicon-based technology. The advantage of the utilized molecular materials over silicon lie in the processibility, i.e. they can be fabricated as flexible substrates, for example. Nevertheless, this example does not permit a dramatic reduction in size to the overall device dimensions.

Silicon industry is currently based on a so-called "top-down" strategy. Herein, small features are developed by processing larger structures. To exemplify this—a table is made by taking a tree, trimming it and furnish it to the desired needs. Transistors, for instance, are etched into silicon substrates using resist and light. However, the increasing demand for miniaturization and densification represents a major challenge for industry, which entered the red zone, that is, reaching the technological limits of current fabrication techniques. Hence, the goal in molecular electronics is slightly different, namely, molecular building blocks, pre-designed for the "bottom-up" approach with specific features and properties, instead of the present solid-state electronic devices which were fabricated from the "top-down" approach using lithographic technologies. Hereby, the prospects of the "bottom-up" approach come into play, since this method implies the *implementation of functionality*, e.g. electron storage, into small features, such as molecules. From this starting point, the self-assembly of such molecules gives rise to higher-order structures, e.g. transistors. The real capacity of such processes comes from thermodynamics. In general, self-assembly is thermodynamically favored ($-\Delta S + \Delta H$), i.e. it is enthalpically favored that the individual components interact to form some organized structures. Thus, "bottom-up" processes are the method of choice to trigger self-assembly in natural systems. In other words, all systems in nature are constructed via "bottom-up". Going back to our analogy of constructing a table, molecular entities would form lipid bilayer membranes and associate into ordered cells, etc. Human beings are the only creatures, which imply "top-down" methods, while nature almost exclusively assembles via "bottom-up". Therefore, one may consider the "bottom-up" approach to be bio-inspired, that is, a natural way of fabrication. Obviously such paradigms influence the use of nanotechnology, and molecular electronics, in particular.

In this context, our ultimate goal is to gain control over specific interactions at the molecular level. Once this goal is achieved, the diversity of self-assembly processes will drive the advances in future technology of electronics. To date, several self-assembled structures, in concert with traditional silicon platforms, have already been made.

A simple play of thoughts might put the enthusiasm and advantages of the "bottom-up" approach into sizeable numbers and visualize its efficiency when compared to the "top-down" method. Ultimately, the structural diversity might lead to more effective molecules with optimal functionality for each application. In that sense, let us consider one mole of molecular switches, with a weight of about 450 g. It can be synthesized in small reactors, e.g. ordinary laboratory flasks with a volume of a couple of liters. Hence, one mole molecular switches would contain 6×10^{23} molecules, which would be more than the combined number of all transistors ever made in the history of the world. Nevertheless, addressing each of these 6×10^{23} molecules is far from reality. The extremely large numbers of switches present in a small flask, displays the potential of molecular electronics in future computing development.

Similarly, we can consider memory, namely the number of bytes of information needed for certain applications. For instance, a color photo is coded by 10^5 bytes, an average book by 10^6 bytes, the genetic code by 10^{10} bytes, the human brain by 10^{13} bytes and the total human culture by 10^{20} bytes. How about employing a mole of bytes? Such a concept is simply striking. Implicit is the task of accessing all these molecules in the required time frame. Still, the possibility of using molecules for such purposes is what renders molecular electronics such an aspiring and powerful tool.

Another example is a Pentium®4 chip. Forty-two million transistors are arranged on a few cm^2 of substrate. Replacing these transistors by organic molecules opens new promises. In particular, an organic molecule that bears source, drain and gate functionality would be 1–3 nm in size. In other words, 10^{14} transistors would fit on a cm^2. Relative to the Pentium®4 technology, this would resemble a 10^6–10^7 increase in density. To date it is, however, impossible to attach more than a single lead to each of those molecules. Nevertheless, simply comparing the size is fascinating. Molecular electronics is still in its infancy and there are numerous obstacles that need to be overcome prior to the realization of a molecular computer. For instance, each of the 42 million transistors on a Pentium®4 chip are addressable and connected to a power supply. Even if it might be easy to be synthesized in large quantities, arrangements on a surface are not trivial. This is even without considering the biggest challenge of all, that is, addressing them on an individual basis. The advantages that quantities and size offer rapidly vanish when throwing micron-resolution of lithographic methods into the game. Nonetheless, the field is rapidly developing with some companies already succeeding in addressing nanoscopic via microscopic techniques. The knowledge of precise arrangement on the nanoscopic scale is, however, scarce. In summary, withdrawing from a silicon-based industry requires a thorough understanding of the serious challenges of molecular electronic device design.

Chemistry alone is not the answer! We are too much concerned with electronic wiring and circuitry. We do not recognize that the beginning of molecular electronics starts with the hybrid approach. The latter implies the addition of molecular systems to present silicon input/output platforms and, hence, achieving viable and competitive computation or memory architectures. Traditionally, chemists have proposed using "cocktails" of molecules and moving them randomly for "computing systems" without realistically interfacing strategies. In other words, putting a photon into a system and harvesting an electron out is a fine laboratory experiment. However, the construction of a device array based upon this typical chemical experiment is wholly impractical.

1.1 Present Technology

An insatiable desire for consumer electronics—for entertainment and communications—has fueled at unparalleled rates the production of smaller, faster, and more powerful logic, memory, display and imaging systems. Indisputably for the industrial progress are the scientific and technological advancements across various fields brought by semiconductor industry to fulfill the customer's aspirations. Still, we should be aware of the core of semiconductor industry's technology, which definitely is the transistor and its supporting electronics which make it possible to utilize. The transistor with two distinguishable switching states (on/off or high/low) is the basic device for computational logic, memory and gain. It implements a boost in power to overcome the signal reduction due to the resistance of the wires upon signal propagation. The technological progress in transistor technology is commonly reflected by "Moore's Law". The latter hypothesizes that the number of transistors per integrated circuit will be doubled on a scale of 18–24 months. This prediction was made by the Intel founder Gordon Moore in a paper from 1965 with the intriguing title "Cramming more components onto integrated circuits" [8]. Nevertheless, this was a prediction rather than a "Law" in a scientific sense. Moore did not believe that this prediction would hold far beyond 1975. However, the exponentially increasing rate of circuit densification has continued into the present times (Fig. 1.1). For instance, in the year 2000, with the introduction of the Pentium®4 processor by Intel with 42 million transistors, an engineering masterpiece was achieved. Nowadays, 8 years later, Intel's most recent Penryn® processor chip already contains 410 million (!) transistors [9]. Thus, the steady development of new fabrication techniques, materials, and processing technologies is responsible for the validity of Moore's Law. Moreover, it demonstrates the commitment toward industrial development by hundreds of thousands of scientists and engineers worldwide.

1.2 Limitations of Present Technology

As one may have expected, theoreticians have gathered numerous possibilities to verify Moore's Law. For instance, they have announced the end of the silicon's densification (silicon brick wall). Similarly, they have predicted the point at which silicon-based devices may not be further reduced. All these theoretical predictions had, however, to be continually revised. This is because the basic models behind these theses was the inability to process ever smaller structures in silicon. In that sense, all predictions were based on technological limitations rather than upon a physical science barrier. It might be serendipitous that engineers have overseen the theoretical predictions hinting to the end of silicon. Instead, they continued to build smaller devices and denser integrated circuits.

1.2 Limitations of Present Technology

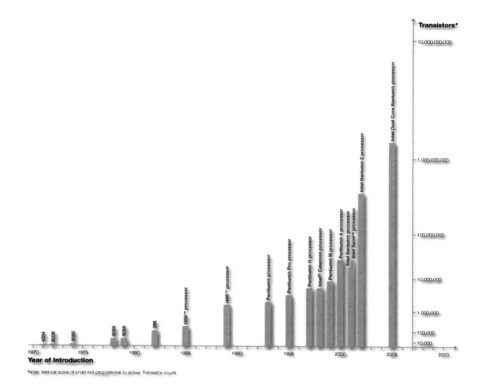

Fig. 1.1 The number of transistors in a logic chip has increased exponentially since 1972 (Intel Data)

These days, the technological limits of the lithographic techniques used to create the circuitry on the wafers go along with the end of the silicon era. However, technologies such as e-beam lithography, extreme ultraviolet lithography (EUV) [10], and X-ray lithography are already commercially in use. They allow standard processing in the sub-100 nm regime down to 45 nm (Intel Penryn® chip) [9].

Apart from these promising developments, we should still bear in mind yet another barrier being approached, that is silicon as a material. In fact, it is not anymore a technological barrier rather than a fundamental physical feature. It implies silicon as a material and is hard to overcome by engineering efforts. For instance, charge leakage arises as a major problem, when the insulating silicon oxide layers were less than three silicon atoms thick. Commercially, such a density was already reached in 2004 [11]. Moreover, quantum confinement imposes serious challenges, as the band structure diminishes at a size domain below these regions. This significantly impacts the electrical and conduction properties of silicon and completely changes the performance of the device. Notable, such limits cannot be overcome by technological breakthroughs. Small materials modifications such as the use of silicon nitride, for example, might lead to a slight deceleration in reaching the physical science barrier of silicon. This is, however, far future at this point.

Summarizing, new technologies are certainly of great interest to the semiconductor industry. Despite the above mentioned limitations, there are several more problems in the chip manufacturing processes regarding financial and environmental issues. Thus, to address at least one of these would certainly be revolutionary. In that sense, for new technologies there are many straps to start to pull at. Certainly, a completely new platform replacing silicon would probably be the best solution. Maybe molecular electronics will provide the right way out.

References

1. Aviram A, Ratner M (eds) (1998) Molecular electronics: science and technology, vol 852. Annul New York Acad Sci
2. Joachim C, Roth S (1997) Atomic and molecular wires, vol 341. NATO Applied Sciences, Kluwer, Boston
3. Tour JM (2000) Acc Chem Res 33:791
4. Joachim C, Gimzewski JK, Aviram A (2000) Nature 408:541
5. Overton R (2000) Wired 8:242
6. Service RF (2001) Science Washington DC US 294:2442
7. Petty MC, Bryce MR, Bloor D (1995) Introduction to molecular electronics. Oxford University Press, New York
8. Moore GE (1965) Electronics 38:8
9. http://www.blogs.intel.com/technology/2007/04/penryn.p. Technology@intel blog, 2007
10. Hand A (2001) Semicond Sci Technol 24:15
11. Packan P (1999) Science Washington DC US 285:2079

Chapter 2
Motivation—Focusing on Molecular Wires

As envisioned above, the rapidly emerging field of molecular electronics offered major incentives to device design and the use of molecular wires [1, 2]. The key idea of using molecules as wires stems from consumer electronics. Let us consider on a journey taking us inside of electronic devices:

Macroscopic wires, namely ordinary metallic wiring, which pass the electron flow in refrigerators, televisions, stereos, computers, household lightning, etc., measure approximately 1 cm in diameter. Going beyond that first level of wiring much smaller wires (approx. 1 mm in diameter) are found on printed circuit boards as they connect smaller components, e.g. resistors, logic chips, rheostats, etc. The next step takes us inside logic chips, for instance. Here, we will come across wires, 10th of a μm wide, which are used to connect solid-state transistors carved out of silicon only. Connecting thousands of such transistors allows performing logic operations. Considering current technologies, this would be pretty much the end of our journey. Reaching beyond this obliges us to overcome the limits of present semiconductor manufacturing methods. Molecular wires would constitute a potential solution. Recent breakthroughs have produced molecular-scale wires, ranging in length from 1 to 100 nm and width from 0.3 nm on up.

In the commercial technology of 2004, the copper wires in Intel's Pentium®4 logic chip are produced in their newest 300 mm wafer fabrication facility in Ireland, and are 90 nm wide [3]. The use of strained silicon [4] is one of several approaches tested to modify present silicon-based processes to meet the demands of the development roadmap. Now, considering a typical molecular wire, investigated in our lab with a width of 0.4 nm and a length of 2.5 nm, see Fig. 2.1. Compared to the Pentium®4 chip 300 of such molecules, side-by-side, would span the 90 nm metal line.

Furthermore, we can tune the physical properties of such a wire in the same way as we can change the raw material used to make it. Thus, the small size, the synthetic diversity, the efficient synthesis of macroscopic amounts in small

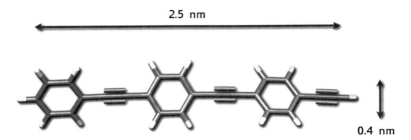

Fig. 2.1 The dimensions of a representative example of a molecular wire are calculated to be 0.4 nm in width and 2.5 nm in length using DFT calculations

reactors are reason enough to prosecute molecular wire research[1] and provide a rational motivation for this thesis: Characterizing several classes of molecular wires.

References

1. Tour JM (2003) Molecular electronics: commercial insights, chemistry, devices, architecture, and programming. World Scientific Publishing, River Edge, New Jersey
2. Tour JM, James DK (2002) Molecular electronic computing architecture. In: Goddard WA, Brenner DW, Lyshevski SE, Iafrate GJ (eds) Handbook of nanoscience, engineering, and technology. CRC Press, Boca Raton, Florida
3. http://www.intel.com/pressroom/archive/releases/20040614corp.htm. Intel Press Release, 2004
4. Singer P (2004) Semiconduct International
5. http://www.itrs.net/Common/2004Update/2004Update.htm. International Technology Roadmap for Semiconductors web pages, 2004.

---

[1] As an example of how far technology has come, molecular electronics is discussed in the "Emerging Research Devices" section of the most recent International Technology Roadmap for Semiconductors [5] and new molecular wires are a large part of the emerging technology.

Part II
Theoretical Concepts

Chapter 3
Concepts of Photoinduced Electron and Energy Transfer Processes Across Molecular Bridges

3.1 Introduction

As we have seen in the previous chapter individual molecules or supramolecular assemblies may perform the functions of electronic devices. Using molecular building blocks to develop electronic circuits mandates the design of specific molecular functionalities. The latter are then the inception to imitate components of an electronic circuit. One of the simplest of these components is a wire. Not surprisingly, the design of "molecular wires" has received a great deal of attention [1, 2]. Despite its simplicity, the definition of this term is rather broad. Some relate it to molecular structures mediating the transport of charge between appropriate donor and acceptor moieties. For instance, one can probe the conduction of molecular wires in break-junction experiments, placing the wire between two tiny gold-rods. In the same way of thinking, the electrodes may be replaced by appropriate donor and acceptor molecules. Generally speaking, we should employ the term "molecular wire" on any molecular structure, which mediates charges between donors and acceptors. In this work, photo- or redox-active organic molecules serve as donors and acceptors. π-conjugated bridges referred to as "molecular wires" link these acceptors and donors.

Molecular wires have been studied under a variety of experimental conditions. Their molecular structures and the nature of the donors and acceptors they have been connected to determine the exact conditions. In fact, the great structural variety of such DONOR–wire–ACCEPTOR systems gives rise to different conduction mechanisms. Thus, numerous detection methods for the electron flow were developed. Some of the available methods include:

- Fast electrochemistry, especially for self-assembled monolayers containing redox-active groups [3–5].
- Conductance measurements in metal-wire-metal junctions ("break junctions") [6–10].
- Photoinduced electron transfer in DONOR–bridge–ACCEPTOR systems [11–13].

Fig. 3.1 Schematic representation of photoinduced electron transfer (*left*) and electronic energy transfer (*right*) in a DONOR-bridge–ACCEPTOR (DBA) system

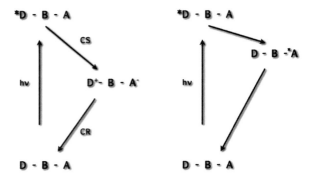

The present work focuses on the last method, namely the investigation of photoinduced electron and energy transfer reactions in organic DONOR(D)–bridge(B)–ACCEPTOR(A) (DBA) conjugates. Nevertheless, comparative assays in light of mechanistic features with electrochemistry and conductance measurements are made to facilitate the overall understanding.

In the aforementioned DBA systems two different mechanisms should be contrasted, namely energy versus electron transfer, where light excitation powers any of the involved reactions. In the following three possibilities for excitation should be discussed: (i) excitation of the donor, (ii) excitation of the bridge and (iii) excitation of the acceptor. Fig. 3.1 schematically depicts one of these possibilities to demonstrate the two different deactivation mechanisms of the light excitation—energy versus electron transfer. In this example, the donor is excited which is most likely to result in electron transfer. Excitation of the acceptor or the bridge, on the other hand, may lead to hole transfer. In principle, electronic energy transfer and electron transfer reactions share a few notable similarities. This, in turn, renders a clear separation rather complicated due to a mutual entanglement of both processes. In most cases, the electron-transfer mechanism follows a chronological sequence of excitation, then energy transfer and finally electron transfer. Herein, those mechanistic aspects will be related to the role of the bridge in the DBA conjugates.

3.2 Electron Transfer Mechanisms

Electron transfer reactions can be described from two different points of view—the classical and the quantum mechanical. Several reviews and publications cover these topics exhaustively [14–18]. Among them, the book by Ulstrup [19] and the excellent review by Weiss et al. [20] constitute real milestones.

From the quantum mechanical perspective, both photoinduced charge separation and charge recombination are radiationless transitions between different,

3.2 Electron Transfer Mechanisms

weakly interacting electronic states in a DONOR/ACCEPTOR conjugate. The finite probability of these processes is given by a golden rule expression:

$$k_{el} = \frac{4\pi}{h}\left(H_{el}^{if}\right)^2 FCWD^{el}. \tag{3.1}$$

Here, H_{el}^{if} is the electronic coupling that links the two states which ultimately interconvert into each other by the electron transfer process. $FCWD^{el}$ is a thermally averaged vibrational Franck–Condon factor, which describes a Franck–Condon weighted density of states. The $FCWD^{el}$ term accounts for the combination of nuclear reorganization energies and driving force effects. $FCWD^{el}$ assists in assessing if the charge-transfer reaction lies in the normal, activationless, or inverted regimes of the classical Marcus theory. In this context, the role of the electronic factor H_{el}^{if} demands for more attention.

When donor–acceptor pairs lack interactions with an intervening medium, e.g. solvent molecules, the electron transfer mechanism is supposed to occur through space. Considering that the electron density of molecular orbitals falls off exponentially, a similar postulate may be formulated for the electronic coupling between donor and acceptor, H_{el}^{if}:

$$H_{el}^{if} = H_{el}^{if}(0)\exp\left[-\frac{\beta}{2}(r_{DA} - r_0)\right]. \tag{3.2}$$

Assuming that $FCWD^{el}$ is distance independent, the electron-transfer rate constants are also expected to decay exponentially as a function of distance:[1]

$$k_{el} = k_{el}(0)\exp[-\beta(r_{DA} - r_0)]. \tag{3.3}$$

Here, r_{DA} is the donor–acceptor distance, $H_{el}^{if}(0)$ is the interaction at constant distance r_0, and β is the "so-called" attenuation factor. In vacuum, values of β are relatively large in the range of 2–5 Å^{-1} [21]. Consequently, at donor–acceptor distances commonly found in molecular dyads, the through space couplings will be negligible.

Fig. 3.2 schematically represents a DBA system, where the donor and the acceptor are separated by a molecular bridge, depicting one of the various structural examples, which will be discussed later in this thesis.

For photoinduced electron-transfer reactions in such DBA systems the bridge must also be taken into consideration. Under these circumstances, two different scenarios have been established depending on the relative energies of the donor, bridge and acceptor levels [11, 12, 22, 23]:

1. If the LUMO of the bridge is significantly higher in energy than the orbitals of the donor, the electron is transferred in a single, coherent step from the donor to

---

[1] For real systems, the distance dependence of the $FCWD^{el}$ term cannot be completely neglected due to the presence of reorganizational energy and the driving force. Thus, a correction for these influences should be done to Eq. 3.3 before applying to experimental results.

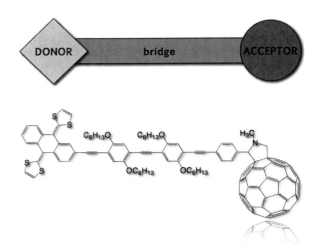

Fig. 3.2 Schematic representation of a DBA system

the acceptor. In such a scenario, the role of the bridge would be limited to a structural function. This kind of mechanism for electron-transfer reactions is often described as *superexchange*.

2. If, on the other hand, the LUMO of the bridge is energetically accessible from the donor orbitals, electron injection from the donor is promoted and the bridge acts as a real intermediate in transferring the electron to the acceptor. This situation is then termed as *charge hopping*.

A comparison between these two types of mechanisms is shown in Fig. 3.3. Demonstrably, these two different mechanisms will impact the experimental results considerably. The distance dependence of electron transfer is certainly one of the features that are impacting the presented DBA systems. However, further theoretical background is needed to understand the different behaviors and the interplay between the two mechanisms.

3.2.1 Superexchange

This mechanism is best understood as coherent tunneling of electrons in a single step, directly from the donor to the acceptor. Nevertheless, the bridge does not solely act as a passive medium. It rather acts as an active linker that modulates the size of the donor–acceptor "through-bond" electronic coupling [18, 24]. These through-bond effects are rationalized in terms of superexchange. Schematically, the coupling between the initial and final states of electron transfer processes originates from the mixing of states. In particular, high-energy *virtual states* of charge-transfer character involving the bridge are considered. The coupling would be negligible in the absence of the bridge. As a result, both superexchange pathways, namely (i) involving donor-to-bridge electron transfer

3.2 Electron Transfer Mechanisms

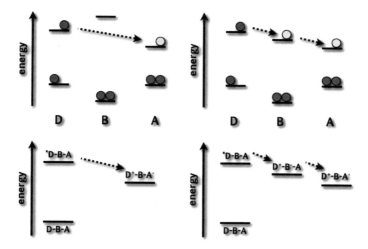

Fig. 3.3 Schematic representation of orbital (*upper*) and state (*lower*) energy diagrams of *superexchange* (*left*) and *hopping* (*right*) mechanisms of photoinduced electron transfer in DBA dyads

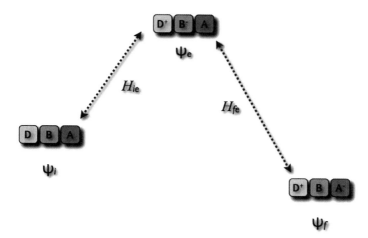

Fig. 3.4 State diagram illustrating *superexchange* interaction between a donor (D) and an acceptor (A) through a simple bridging group (B). Ψ_e is the electron-transfer "virtual" state involving the bridge. For the other symbols, see text

and (ii) bridge-to-acceptor hole transfer virtual states, should be considered [11, 12, 18, 22, 23, 25–29]. However, usually one of the pathways predominates and thus in the following we will focus on the electron transfer pathway. Fig. 3.4 illustrates a state diagram of the superexchange interaction between a donor D and acceptor A through a simple bridging group B. Related to second order perturbation theory the electronic superexchange coupling is expressed by

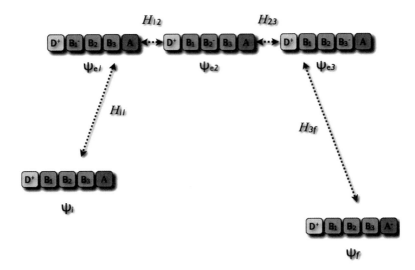

Fig. 3.5 State diagram illustrating *superexchange* interaction between a donor (D) and an acceptor (A) through a modular bridge. ψ_{e1}, ψ_{e2} and ψ_{e3} are local electron-transfer "virtual" states

$$H_{if}^{el} = \frac{H_{ie}H_{fe}}{\Delta E_e}, \tag{3.4}$$

where H_{ie}, H_{fe}, are the donor–bridge and bridge–acceptor coupling elements, respectively (see Fig. 3.4). ΔE_e is the energetic difference between the virtual state and the initial/final state. Importantly, these differences should be taken at that transition state geometry, where the initial and final state have the same energy. Eq. 3.4 points out that the key factor in determining the mediating properties of the bridge is the energy denominator. Easily reducible bridges, i.e. with low ΔE_e values, and therefore relatively low-energy LUMOs, are good electron-transfer superexchange mediators. On the other hand, considering a hole-transfer pathway, a similar argument would hold for easily oxidizable bridges. In the simple schematic representation from Fig. 3.4 only one single virtual state is taken into account. In dyads the situation is, however, often more complicated. As the bridges have mostly a modular structure, constructed from individual weakly interacting units, not only the chemical nature but also the length of the bridge will be relevant to its conducting properties.

Applying these thoughts to Fig. 3.4, leads to an extension of the state diagram as demonstrated in Fig. 3.5. Now, we have to include virtual charge-transfer states localized on each single modular unit. For a three-module bridge, for instance, the perturbation expression for an electron-transfer reaction extends to

$$H_{if}^{el} = \frac{H_{i1}H_{12}H_{23}}{\Delta E_{i1}\Delta E_{i2}\Delta E_{i3}}H_{3f}. \tag{3.5}$$

3.2 Electron Transfer Mechanisms

Thus, for n identical modular units, the donor–acceptor superexchange coupling transforms to

$$H_{if}^{el} = \frac{H_{i1}H_{nf}}{\Delta E_{in}} \left(\frac{H_{12}}{\Delta E}\right)^{n-1}. \tag{3.6}$$

Not surprisingly, for a through-bond superexchange interaction we obtain an exponential dependence of H_{if}^{el} on the number of the modular units in the bridge, i.e. a dependence on the length of the bridge. For linear bridges, this simply transforms into an exponential dependence of H_{if}^{el} on donor–acceptor distance. Hence, we still can make use of Eqs. 3.2 and 3.3 since they characterize the distance dependence of the donor–acceptor coupling through a molecular bridge. For further insight, it is useful to compare Eq. 3.2 with Eq. 3.6. Now, we can understand the meaning of the various terms in Eq. 3.2:

1. r_0 and $H_{if}^{el}(0) = \frac{(H_{i1}H_{1f})}{\Delta E}$ represent the donor–acceptor distance and the effective coupling for a hypothetical bridge with one single module, respectively.
2. The attenuation factor $\beta = 2\ln\left(\frac{H_{12}}{\Delta E}\right)$ becomes a bridge specific parameter and depends on the magnitude of the coupling between adjacent modular sites and the energy of the electron- (or hole-) transfer states localized on each module.

This theoretically established distance dependence has been verified in several homogeneous series of dyads containing modular organic linkers of variable length [22, 30–32].

Concluding, in the superexchange regime, the bridge plays an active role in opening a path for electronic coupling between donor and acceptor. On that basis we finally reach the first definition of "molecular wires". In that sense, bridges may be considered as "molecular wires" although their properties drastically differ from those of a conventional ohmic conductor. The distance dependence of electron transfer probability, as seen from the aforementioned discussion, in organic wire-like molecules offers novel approaches to effectively conduct electrons while omitting *Ohm's Law*, even at room temperature. Yet, *superexchange coupling* may suffer from *charge hopping*: As a matter of fact both mechanisms should be carefully defined before implementing the concepts of lossless transport of electrons through molecular bridges.

3.2.2 Charge Hopping

The *charge hopping* regime brings the molecular bridge by far closer to the role of a real (molecular) wire. In this regime, the energy levels of the bridge allow a sequential donor-to-bridge and bridge-to-acceptor electron transfer. A schematic representation is shown in Figs. 3.3 and 3.6. Inspecting the latter scenario, we can divide the charge transfer into three different processes with three different rate constants:

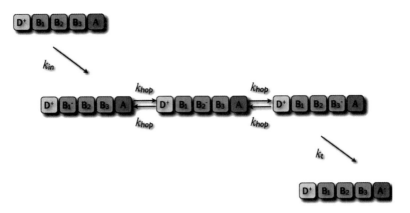

Fig. 3.6 *Charge hopping* in a donor–bridge–acceptor system involving a modular bridge

1. Charge injection from the donor into the bridge; occurring with k_{in}.
2. Hopping among (ideally) isoenergetic charge-transfer states of adjacent bridge subunits; reflected by k_{hop}.
3. Trapping of the electron at the acceptor; k_t.

Assuming, that charge hopping involving different bridge units is a diffusive process a simple treatment leads to the overall rate constant [33]

$$k = k_{hop}\left(\frac{r_{DA}}{r_L}\right)^{-2} = k_{hop}N^{-2}. \tag{3.7}$$

Here, r_{DA} is again the total donor–acceptor distance; r_L, the length of each modular unit of the bridge; and N, the number of units. Eq. 3.7 predicts that the distance dependence of charge transfer rates should be relatively weak in the hopping regime. When compared to ohmic conductors with a r_{DA}^{-1} distance dependence, the bridge will undoubtedly behave differently. In the hopping regime the bridge will actively participate in mediating charges over long distances. In this context, the term "molecular wire" is more suitable for this regime.

To meet the requirements for effective *charge hopping*, namely low reduction or oxidation potentials and fine tuning of the energy gradients, the chemical nature of the bridge is considered to play a major role. Experimentally, in donor–acceptor dyads the charge-hopping regime is less common, unless we are dealing with potent electron donors, such as TTF-derivatives, etc. and highly conjugated bridges [32], which is the key-issue of this thesis. Not until these conditions are met, it is possible to obtain superficial distance dependencies with very low attenuation factors, β, far below 0.5 Å^{-1}, when the data are plotted according to an exponential law such as in Eq. 3.3 [13, 30, 31, 34]. A worth mentioning example, not related to DBA dyads, is the charge hopping in DNA where efficient hole transfer is prominent [35].

3.2.3 Interplay of Mechanisms

The general concepts as they were derived in the previous section apply only to theoretical models and/or idealized systems. In real systems, both superexchange and hopping mechanisms through modular bridges require a more sophisticated treatment. In fact, both mechanisms are omnipresent in charge-transfer reactions and interplay with each other.

Superexchange in a modular system assumes, for instance, that the bridge energy levels are independent on the number of the bridge units (Fig. 3.5). In a strict sense, such an assumption is only applicable for non-interacting bridge units. But, with $H_{12} = 0$ superexchange-mediated electron transfer would not take place. In reality, $H_{12} \neq 0$ and an increase in bridge length causes a more or less pronounced lowering in the bridge energy levels. Still, structural examples are present, with a very low dependence of the oxidation potential, i.e. bridge energy levels, on the length of the bridge. This will be discussed in more detail at a later point.

Several consequences as they stem from such a non-ideal behavior of the bridge energy levels as a function of bridge units impact both mechanisms. First of all, the electronic coupling between the bridge and the donor/acceptor sites will change when ΔE is not constant in Eq. 3.6. This, in turn, affects the exponential distance dependence of the transfer rates, i.e. the attenuation factor β. More importantly, an increase in bridge length can even cause a switch in mechanism from superexchange to hopping. Especially, when lowering the bridge levels at larger bridges below those of the donor levels. Such a scenario may also occur when changing the temperature. Under these conditions, at any given bridge length or temperature an abrupt mechanism change is expected to occur. Often, a sudden increase in rate constants accompanies a transition from strong to weak distance dependence. Numerous examples of this behavior have been already reported [13, 36] and several new cases have been elaborated in this thesis.

At this point, we have to consider yet another very important process that is likely to go hand in hand with photoinduced electron transfer reactions—the electronic energy transfer.

3.3 Electronic Energy Transfer

In a supramolecular system, electronic energy transfer, as depicted in Fig. 3.1, can be viewed as a radiationless transition between two locally, electronically excited states. Similarly to the electron transfer described above, we deal with two different states. Nevertheless, these states are local excitations lacking any charge transfer.

$$^\star D - B - A \longrightarrow D - B - A^\star \qquad (3.8)$$

Thus, the rate constant for the energy transfer process is given by a golden rule expression:

$$k_{en} = \frac{4\pi}{h}\left(H_{if}^{en}\right)^2 FCWD^{en}. \qquad (3.9)$$

Again, H_{if}^{en} is the electronic coupling between the two excited states interconverted by the energy transfer process and $FCWD^{en}$ is an appropriate Franck–Condon factor.

Equally to what has been done for the electron transfer processes, the Franck–Condon factor is derived from either quantum mechanical [37–39] or classical [40] expressions. From a quantum mechanical point of view, this factor is the thermally-averaged sum of vibrational overlap integrals. They represent the distribution of transition probabilities over several isoenergetic virtual transitions, i.e. from $^\star D$ to D and $^\star A$ to A, in the two molecular units. In other words, it describes the transition probability between two or more vibrational states. In classical terms, it accounts for effects of nuclear reorganization energies and the energy gradients as they both impact the rate constant. Experimental information is gathered by evaluating the overlap integral between the emission spectrum of the donor and the absorption spectrum of the acceptor.

On the other hand, the electronic factor, H_{if}^{en}, is a two-electron matrix element, involving the HOMOs and LUMOs of the energy-donating center and the energy-accepting center, respectively. Following standard photophysics [41], this factor consists of two additive terms—a *coulombic* and an *exchange* term. These two terms, in turn, depend on several photophysical/structural parameters, e.g. donor–acceptor distances, spins of ground and excited states, spin–orbit couplings, etc. As a consequence, we can identify two main energy transfer pathways, namely the *coulombic* or *resonance* and the *exchange* energy transfer mechanism. Both are schematically represented in Fig. 3.7.

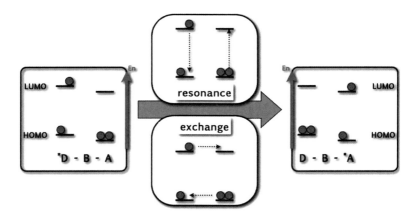

Fig. 3.7 Pictorial representation of *resonance* (*coulombic*) and *exchange* energy transfer mechanisms

3.3.1 Coulombic Energy Transfer

The *coulombic* is also known as *resonance*, *dipole–dipole*, or *Förster-type* mechanism [42]. This type of energy-transfer process does not require physical contact (i.e. exchange interaction etc.) between donor and acceptor. The most important factor within the coulombic interaction is the *dipole–dipole* term, that obeys the same selection rules as the corresponding electric dipole transitions of the two exchange partners. Thus, if the oscillator strength of the radiative transitions connecting the ground and excited state of each unit is high, couloumbic energy transfer is most probable. In order to correlate the rate constant for coulombic energy transfer with spectroscopic and photophysical properties of the two molecular components, we need to evaluate the classical *Förster* formula, given by

$$k_{en}^{coul} = 1.25 \times 10^{17} \frac{\Phi_D}{n^4 \tau_D r_{DA}^6} \int_0^\infty F_D(\bar{\nu}) \varepsilon_A \frac{d\bar{\nu}}{\bar{\nu}^4}, \quad (3.10)$$

where Φ_D is the quantum yield of the donor emission, n is the solvent refractive index, τ_D is the lifetime of the donor emission, r_{DA} is the nm-distance between donor and acceptor, F_D is the emission spectrum of the donor (in wavenumbers and normalized to unity), and ε_A is the decadic molar extinction coefficient of the acceptor. Resulting from this equation, energy transfer may efficiently occur over distances of up to 50 Å due to the $\left(\frac{1}{r_{DA}^6}\right)$ distance dependence of the rate constant, when a good spectral overlap integral and appropriate photophysical parameters are guaranteed. For that reason, *coulombic* energy transfer is also referred to as long-range energy transfer. As a leading example the singlet–singlet energy transfer mechanism should be regarded:

$$^\star D(S_1) - B - A(S_0) \rightarrow D(S_0) - B - A(S_1)^\star. \quad (3.11)$$

Examples of efficient energy transfer in organic DBA systems will be provided later in the course of this thesis, especially dealing with singlet–singlet energy transfer. In highly conjugated molecular bridges the lifetime of the singlet excited states is very short, so that coulombic energy transfer is less exhibited [43, 44].

3.3.2 Exchange Energy Transfer

In contrast to the *coulombic* energy transfer mechanism, the *exchange* energy transfer mechanism (also called *Dexter*-type) [42] occurs typically at short range and is limited to orbital overlap, in other words, physical contact between the donor and the acceptor. A simple way to visualize exchange interactions, as demonstrated in Fig. 3.7, is to imagine a simultaneous exchange of two electrons. For this process, of course, spin conservation must be obeyed in the reacting pair as a whole. Consequently, exchange mechanisms are applicable in those cases, where the excited states are involved that are spin-forbidden. Triplet–triplet energy transfer meets these requirements and evolved as a benchmark example for efficient exchange:

$$^\star D(T_1) - B - A(S_0) \rightarrow D(S_0) - B - A(T_1)^\star. \tag{3.12}$$

Exchange energy transfer from the lowest spin forbidden excited state is expected for singlet–triplet intersystem-crossing (ISC) reactions presented further in the following parts of this thesis.

Similarly to the discussion above with respect to electron-transfer processes, the rate constant of *exchange* energy transfer reflects the nature of the bridge. Arguments that were brought forward for superexchange hold also for a mixing with electronically excited states that are localized on the bridge, which is expected to effectively mediate the donor–acceptor exchange interaction. Again, bridges with low-energy excited states will be particularly efficient. A direct one-step may convert to a sequence of donor-to-bridge and bridge-to-acceptor energy migration steps if the electronically excited states of the bridge drop below the energy of the donor. In long, modular bridges the exchange interaction is expected to fall off exponentially as function of distance [45–47].

Interestingly, in a homogeneous series of dyads where triplet energy transfer (Eq. 3.12), electron transfer (Eq. 3.13), and hole transfer (Eq. 3.14) processes

$$D^- - B - A \rightarrow D - B - A^- \tag{3.13}$$

$$D^+ - B - A \rightarrow D - B - A^+ \tag{3.14}$$

could be studied as a function of distance across the same saturated organic bridges, the value of the attenuation factor β (see Eq. 3.2) for energy transfer was found to be the sum of those for electron and hole transfer [48]. This supports the simple notion that exchange energy transfer is equivalent to a simultaneous double electron transfer. To this end, the electronic matrix element is proportional to the product of those for the corresponding electron and hole transfer processes.

References

1. Jortner J, Ratner M (eds) (1997) Molecular electronics. Blackwell, London
2. Joachim C, Gimzewski JK, Aviram A (2000) Nature 408:541
3. Craeger S, Yu CJ, Bamad C, O'Connor S, MacLean T, Lam E, Chong Y, Olsen GT, Luo J, Gozin M, Kayyem JF (1999) J Am Chem Soc 121:1059
4. Sachs SB, Dudek SP, Hsung RP, Sita LR, Smalley JF, Newton MD, Feldberg SW, Chidsey CED (1997) J Am Chem Soc 10:563
5. Sykes HD, Smalley JF, Dudek SP, Cook AR, Newton MD, Chidsey CED, Felberg SW (2001) Science Washington DC US 291:1519
6. Donhauser ZJ, Mantooth BA, Kelly KF, Bumm LA, Monnell JD, Stapleton JJ, Price DW Jr., Rawlett AM, Allara DL, Tour JM, Weiss PS (2001) Science Washington DC US 292:2303

References

7. Chen J, Reed MA, Rawlett AM, Tour JM (1999) Science Washington DC US 286:1550
8. Tour JM, Rawlett AM, Kozaki M, Yao Y, Jagessar RC, Dirk SM, Price DW, Reed MA, Zhou C-W, Chen J, Wang W, Campbell I (2001) Chem Eur J 7:5118
9. Haag R, Rampi MA, Holmlin RE, Whitesides GM (1999) J Am Chem Soc 121:7895
10. Holmlin EH, Ismagilov RF, Haag R, Mujica V, Ratner MA, Rampi MA, Whitesides GM (2001) Angew Chem Int Ed 40:2316
11. Scandola F, Chiorboli C, Indelli MT, Rampi MA (2001) Covalently linked systems containing metal complexes. In: Balzani V (eds) Electron transfer in chemistry, vol III, chap. 2.1. Wiley, Weinheim, p 337
12. De Cola L, Belser P (2001) Photonic wires containing metal complexes. In: Balzani V (eds), Electron transfer in chemistry, vol V, chap. 3. Wiley, Weinheim, p 97
13. Davis WB, Svec WA, Ratner MA, Wasielewski MR (1998) Nature 396:60
14. Marcus RA (1964) Annu Rev Phys Chem 15:155
15. Sutin N (1983) Prog Inorg Chem 30:441
16. Marcus RA and Sutin N (1985) Biochim Biophys Acta 811:265
17. Jortner J (1976) J Chem Phys 64:4860
18. Newton MD (1991) Chem Rev Washington DC US 91:767
19. Ulstrup J (eds) (1979) Charge transfer processes in condensed media. Springer, Berlin, Heidelberg, New York
20. Weiss EA (eds) (2005). Top Curr Chem, vol 257. Springer, Berlin, Heidelberg
21. Newton MD (1991) Chem Rev Washington DC US 91:767
22. Paddon-Row MN (2001). Covalently linked systems based on organic components. In: Balzani V (eds) Electron transfer in chemistry. vol III, chap. 2.3. Wiley, Weinheim, p 179
23. Jortner J, Bixon M, Langenbacher T, Michel-Beyerle ME (1998) Proc Natl Acad Sci USA 95:759
24. Oevering H, Verhoeven JW, Paddon MN-Row, Warman JM (1989) Tetrahedron 45:4751
25. Halpern J and Orgel LE (1960) Faraday Discuss 29:32
26. McConnell HM (1961) J. Chem. Phys., 35:508
27. Mayoh B, and Day P (1974) Dalton Trans 8:846
28. Miller JR, and Beitz JV (1981) J Chem Phys 74:6746
29. Scandola F, Argazzi R, Bignozzi CA, Chiorboli C, Indelli MT, Rampi MA (1993) Coord Chem Rev 125:283
30. Atienza C, Martfn N, Wielopolski M, Haworth N, Clark T, Guldi DM (2006) Chem Commun Cambridge UK 30:3202
31. Atienza-Castellanos C, Wielopolski M, Guldi DM, Pol Cvd, Bryce MR, Filippone S (2007) N Martin Chem Commun Cambridge UK 48:5164
32. Wielopolski M, Atienza-Castellanos C, Clark T, Guldi DM, Martfn N (2008) Chem- Eur J 14:6379
33. Meggers E, Michel-Beyerle ME, Giese B (1998) J Am Chem Soc 12:950
34. Atienza C, Insuasty B, Seoane C, Martfn N, Ramey J, Guldi DM (2005) J Mater Chem 15:124
35. Lewis FD (2001) Electron transfer and charge transport processes. In: Balzani V (ed) Electron transfer in chemistry, vol III, Chap. 1.5. Wiley, Weinheim, p 105
36. Weiss EA, Ahrens MJ, Sinks LE, Gusev AV, Ratner MA, Wasielewsi MR (2004) J Am Chem Soc 126:5577
37. Orlandi G, Monti S, Barigelletti F, Balzani V (1980) Chem Phys 52:313
38. Murtaza Z, Zipp AP, World LA, Graff D, Jones WE Jr, Bates WD, Meyer TJ (1991) J Am Chem Soc 113:5113
39. Naqvi KR and Steel C (1993) Spectrosc Lett 26:1761
40. Balzani V, Bolletta F, Scandola F (1980) J Am Chem Soc 102:2152
41. Lamola AA (1969) In: Lamola AA, Turro NJ (eds) Energy transfer and organic photochemistry. Wiley, New York
42. Turro NJ (1978) Modern molecular photochemistry. Benjamin Cummings, Menlo Park

43. Pol Cvd, Bryce MR, Wielopolski M, Atienza-Castellanos C, Guldi DM, Filippone S, Martín N (2007) J Org Chem 72:6662
44. Gnichwitz J-F, Wielopolski M, Hartnagel K, Hartnagel U, Guldi DM, Hirsch A (2008) J Am Chem Soc 130:8491
45. Scholes GD, Ghiggino KP, Oliver AM, Paddon-Row MN (1993) J Phys Chem A 11:871
46. Oevering H, Verhoeven JW, Paddon-Row MN, Cotsaris E, Hush NS (1988) Chem Phys Lett 143:488
47. Closs GL, Piotrowiak P, McInnis JM, Fleming GR (1988) J Am Chem Soc 110:2652
48. Closs GL, Johnson MD, Miller JR, Piotrowiak P (1989) J Am Chem Soc 111:3751

Chapter 4
Molecule-Assisted Transport of Charges and Energy Across Donor–Wire–Acceptor Junctions

The previous sections demonstrated that charge-transfer and energy-transfer processes are characterized by many different parameters. We cannot limit ourselves to the molecular building blocks of a system. Instead, the entire supramolecular structure should be analyzed as a whole system. Only this assists in gathering a sufficient understanding of the interplay between the components. In particular, a close inspection reveals that the energetics and the relationship of the energy-levels of donor, bridge and acceptor govern the energy/charge transfer properties of these systems. However, the key roles in these transport processes are played by the molecular bridges connecting the donor with the acceptor. In the following we survey these processes with particular emphasis on the function of the bridges, or, in other words, on their *molecular wire* properties.

In this part of the thesis, we will combine the theoretical aspects of the previous section with experimental aspects of charge/energy transport and focus on the molecular wire-like transport from the mechanistic point of view.

Regarding the correct terminology, there are several references and examples of the term "molecular wire". In some cases, it describes a system with a very specific behavior. In others, it simply refers to the structural features of the molecule under consideration. Thus, finding a clear definition is a rather difficult task. In 1998 an attempt was made by Emberly and Kirczenow [1] and a *molecular wire* has been defined as "a molecule between two reservoirs of electrons". Nitzan and Ratner, on the other hand, called it "a molecule that conducts electrical current between two electrodes" [2]. Most appropriate with respect to the topic of this thesis, we should stick to a rather restricted definition by Wasielewski, which classifies a *molecular wire* as "a device that conducts in a regime, wherein the distance dependence (of electron transfer) may be very weak" [3].

More important than the definition is the desired function of a molecular wire system. According to our discussions of the previous section, we now can define certain criteria that molecules have to meet to act as *molecular wires*. In this context it is possible to point out the desired features that make molecules potential candidates for the electronic components:

1. Structural tunability, due to a sheer unlimited number of synthetic possibilities allowing higher control and versatility relative to metals and semimetals.
2. Orbitals of predisposed shape direct the movement of charge carriers within the molecule.
3. Electron density distribution in the molecules presides over spin preservation preventing the redistribution of spin to other electrons in the system in form of polarization.

Energy and charge transport phenomena along a molecular wire are closely related, despite the fact, that energy transfer is represented as a motion of electron/hole pairs (Frenkel exciton or boson) and charge transfer as a motion of a single hole or electron (a fermion). When considering transport processes, we can consider molecular wires as Hückel-like or tight-binding systems, dominated by nearest-neighbor contacts. A thorough quantum chemical examination of Hückel-like systems can be found elsewhere [4–6].

Now, as we have defined wire-like transport by a motion that is assisted by molecular bridges, we may proceed to the mechanistic point of view. In particular, we will contrast theoretical results with experimental data for molecular wire-like behavior in regard to the transfer of electronic charge and/or energy. Intramolecular electron-transfer (ET) rate constants characterize the charge transport in DBA conjugates and in electronic transport junction we can apply the word "conductance".

In this thesis, we will focus on molecules as individual wires. It is, nevertheless, important to point out, that electron and energy transport in higher dimensional molecular materials (such as polymers or molecular crystals) is closely interconnected. Many of the mechanisms that have been discussed in this thesis, such as coherent tunneling, thermal hopping, Förster transfer, etc., are present in both the single-wire systems and in molecular materials.

4.1 Mechanisms of Charge Transfer Through Molecular Wires

Intrinsically, long-distance charge transfer (CT) is a nonadiabatic process. The rate of CT is determined by a combination of distance-dependent tunneling mechanism and distance-dependent incoherent transport that are strong and weak, respectively [7, 8]. As already discussed the tunneling obeys a superexchange mechanism, where electrons or holes are transferred from donor to acceptor through an energetically isolated bridge. Herein, the bridge orbitals are considered solely as a coupling medium with no nuclear motion along the bridge [9–11]. On the other hand, incoherent or sequential CT involves real intermediate states that couple to internal nuclear motions of the bridge and the surrounding medium. Such states are energetically accessible and may change their geometry [12]. This mechanism is defined as thermally activated incoherent hopping [3, 13].

4.1 Mechanisms of Charge Transfer Through Molecular Wires

Accordingly, the attenuation factor β is used to describe the quality of a molecular wire, owing to the fact that it describes the decay of the CT rate constant k as a function of distance, r_{DA},

$$k = k_0 \exp^{-\beta r_{DA}}. \qquad (4.1)$$

Furthermore, we can use β to predict whether a DBA system should be considered to display wire-like behavior at all. Eq. 4.1 predicts that at small β the charges might be transferred over larger distances. Importantly, such a definition of β only applies to exponentially decaying processes. In addition, the limit of very small β equals to band transport, a so-called "π-way through" which the electrons can travel coherently [14]. A shallow distance dependence may also result from a series of short-range tunneling events. The latter is equivalent to incoherent hopping which, per se, does not follow an exponential decay when considering increasing length. Implicit is that β is an intrinsic factor. This holds for the entire wire system rather than only for the molecule that is providing the linkage between donor and acceptor. In this respect, there is no fundamental difference between donor and acceptor from metallic contacts linked to a spacer.

Interestingly, β values for identical bridge units differ significantly. β reflects, for example, the sensitivity to the surrounding environment of transport. Length dependence is one of these parameters. Exemplary, β values determined for bridge units in solution and through self-assembled monolayers (SAMs) of organic thiols on the surfaces of metal electrodes (Ag, Au and Hg) are listed:

- $\beta = 0.6$–1.2 Å^{-1} for *alkanes* [15–21].
- $\beta = 0.32$–0.66 Å^{-1} for *oligophenylenes* [22–26].
- $\beta = 0.01$–0.5 Å^{-1} for *oligo(phenylene ethynylene)s* (*o*PEs) and *oligo(phenylenevinylene)s* (*o*PVs) [27–31].
- $\beta = 0.04$–0.2 Å^{-1} for *oligoenes* [28, 32, 33].
- $\beta = 0.04$–0.17 Å^{-1} for *oligoynes* [34–36].

4.1.1 Superexchange Charge Transfer in Molecular Wires

Now upon establishing the theoretical framework for the description of molecular-wire behavior, we should analyze the transport phenomena as a function of molecular-wire properties. Firstly, we will discuss the superexchange mechanism. The latter is considered as the main mechanism for efficient electron-transfer within the photosynthetic reaction center [37, 38] and has been studied in various biomimetic systems [39, 40].

In superexchange CT reactions, no charge ever actually resides on the bridge. Those states that the molecule occupies between the time when the electron leaves the donor and arrives at the acceptor are called virtual excitations. The transmission probability of electrons to reach the acceptor in this manner is the superexchange electronic coupling, t_{DA} [7, 41]. Such coupling term is calculated by using a perturbation-theory-based expression. It relates to individual resonance integrals between molecular subunits and the energy gap between the donor and acceptor and the bridge [42]. The model assumes an exponential dependence of t_{DA} on the donor–acceptor distance, r_{DA}. The decay is once more expressed by a β^{SE} parameter:

$$\beta^{SE} = -2\left(\ln\left|\frac{t}{\Delta}\right|\right)/r. \tag{4.2}$$

Here t is the nearest-neighbor transfer integral, Δ is the energy of the bridge orbitals relative to the donor/acceptor and r is the width of one subunit. This expression has been validated under various experiftions [43–45]. Furthermore, it opens up the possibility to calculate the probability of CT via superexchange according to the chemical composition of the bridge.

4.1.2 Sequential Charge Transfer in Molecular Wires

The superexchange coupling decays exponentially with increasing donor–acceptor distance. Now, when the length of the bridge increases the probability of superexchange interaction can become very small. Under these circumstances, if the donor and the acceptor are within $k_B T$ in resonance, the CT rate constant may be dominated by an incoherent term. Several conditions have to be fulfilled for sequential CT. According to Jortner et al. [46] sequential CT takes place when (i) near-resonant charge injection is present, (ii) vibronic overlap exists between ion pair states that are formed while charges move from one bridge site to another, and (iii) vibronic overlap between the ion pair state given by the terminal bridge site and the ultimate acceptor. Importantly, the distance dependence in this case is sensitive to the nature of diffusing charges from the donor to the acceptor. It requires sufficient free energy to mediate the charges from donor to acceptor, and thus the CT rate is given by:

$$\ln k_{CT} \propto -\eta \ln N, \tag{4.3}$$

Here, η is between 1 and 2, N is the number of hopping steps, and the proportionality constant is dependent on the individual tunneling probabilities [14, 46, 47]. As a consequence, long distance incoherent transfer is in general more efficient than the coherent analogue [3, 48]. Obviously at intermediate distances, a competition between superexchange and sequential mechanism arises. This problem has been treated extensively in theory in order to understand both cases [49–51]. Theoretical approaches emanate from the limiting scenario of the following three states:

1. The initial state, where donor, bridge and acceptor are neutral.
2. The intermediate state, whose energy is systematically varied while the donor is oxidized and the bridge reduced.
3. The final state, where the donor is oxidized and the acceptor reduced.

Without going into details, several conclusions should be gathered. First of all, large energy gaps between the initial and intermediate states propagate superexchange mechanisms. They are seemingly activationless. Competition between the superexchange mechanism and sequential mechanism commences when the energy gap approaches the limits of reorganization energies or electronic couplings.

Before, however, proceeding to the results, we should discuss electronic and structural parameters of the given systems as they potentially impact the charge-transfer mechanism.

4.2 Factors that Determine the Charge Transfer Mechanism

4.2.1 Electronic Coupling

The electronic coupling between the components constituting molecular or metallic DBA systems influences both incoherent and superexchange electron transport between donors and acceptors. For the superexchange mechanism, for instance, the most relevant quantity is the donor–acceptor superexchange coupling, t_{DA}. On the other hand, if the bridge levels are energetically accessible, as for the incoherent mechanism, the coupling between the hopping sites will be rate determining.

4.2.1.1 Calculation of Electronic Coupling

The calculation of direct and indirect coupling has been a theoretical challenge. It is mainly the involvement of excited state configurations and the demand for reliable experimental data that renders this difficult. Several concepts have been formulated—all dependent on the particular situation. For instance, the generalized Mulliken Hush (GMH) theory [52, 53] uses parameters that are derived from optical spectra in such cases where the CT state is optically accessible. Alternatively, it may be calculated using methods that afford excited state energies and dipole moments. Most common methods reveal, however, great flaws in handling long-distance ET systems. CT states that bear t_{DA} are not CT states at all but rather real radical ion pair states (RPs). Usually, these RPs posses energies that are on the scale of electron-volts above the neutral ground state. Another characteristic of these RPs is their vanishing transition moments. In fact, they lack any appreciable emission during charge recombination. Acceptable determination of GMH parameters becomes difficult.

Alternative approaches extract the electronic couplings for individual ET processes from calculations of the free energy for reaction and the internal and external reorganization parameters [54]. However, most of the methods depend on available spectroscopic data. One example is the method by Stuchebrukov et al. [55], which employs the Green's function. By modeling the structure of donor and acceptor, they found that ET takes place mostly through a system of overlapping π-orbitals. Then t_{DA} can be expressed in terms of the Green's function of the bridge in the nonorthogonal extended Hückel basis. By rewriting t_{DA} in terms of transition amplitudes, T_i, they were able to derive an expression for the probabilities of "superexchange transitions" between the donor orbitals and the bridge orbitals:

$$t_{DA}(E) = \sum_i \langle A | V | i \rangle T_i. \tag{4.4}$$

V is the strength of the coupling to the bridge and $|i\rangle$ are the orbitals of the bridge. Using Green's function methods they could obtain the amplitudes T_i by solving a system of linear equations,

$$\langle i | V | D \rangle = \sum_j (ES - H^B)_{ij} T_i, \tag{4.5}$$

using conjugate gradient methods [56, 57]. ($S = \langle i | j \rangle$ is the overlap matrix and H^B is the Hamiltonian of the bridge in the extended Hückel basis set $|i\rangle$). Nevertheless, the practical calculation is not as trivial as it sounds. In general, ample experimental data sets on charge migration are needed to validate such calculations.

Needless to say, many extensions of the herein presented methods and different approaches are used to theoretically determine the electronic coupling. Conventional quantum chemistry ranging from fundamental ab initio methods to sophisticated semi-empirical approaches for the calculation of excited state properties is a useful tool to evaluate molecular-wire properties.

4.2.2 Energy Matching

Next, another key step of CT reactions in DBA systems is the charge injection into the bridge. Thus, the evaluation of charge injection energies is crucial when examining the transport properties. Generally speaking, charge injection energies are given by the energies of the frontier orbitals of the molecular bridge relative to the molecular levels of the donor and acceptor or the Fermi energy of the metallic contacts. This description is slightly misleading but very common. Exactly, we should use the state energies of the donor (or the Fermi level) and of the bridge cation and bridge anion. However, the frontier orbital energies are often a reasonable approximation to these multi-electron values. In particular, electron transport across metal-molecule-metal junctions depends strongly on the position

4.2 Factors that Determine the Charge Transfer Mechanism

of the Fermi level of the metal electrodes relative to the LUMO and the HOMO of the molecular bridge. DBA systems can be considered equally.

Concluding the findings from the previous sections, we now call back to mind two probable situations:

1. When the energetic difference between the LUMO and the Fermi level is large, electron transport occurs by superexchange tunneling, i.e. tunneling mediated by interactions between donor and acceptor and unoccupied orbitals of the organic bridges that separate them.
2. If the Fermi level approaches the energy of the orbitals of the molecular bridge, resonant electron transfer may take place—either by hopping or resonant tunneling. In this case the conduction of electrons will occur through the molecular orbitals.

Fan et al. [58] was able to prove these two mechanisms by means of Shear–force-based scanning probe microscopy (SPM) experiments. These exemplify the importance of energy matching when determining the transport mechanism. SPM was complemented by current–voltage (I–V) and current–distance (I–d) measurements. In these investigations, ET processes were tested across SAMs of hexadecanethiol and nitro-substituted compounds, including *oligo*(phenyleneethynylene) (*o*PE) on gold surfaces. In these simple experiments, the hexadecanethiol SAMs gave rise to an exponential current increase as the distance was decreased. Large β values that ranged from 1.3 Å^{-1} to 1.4 Å^{-1} resulted. In the low bias regime, the dependence was nearly independent on the tip bias. The currents increased, however, exponentially with bias in the high-bias regime. All these observations suggest a superexchange mechanism for transport through SAMs. The mismatch between molecular orbitals of the *o*PEs and the Fermi level of gold might be responsible for this trend.

Placing of nitro groups into the *o*PE structures changed the observation. Important is the fact that nitro-substituted *o*PEs contain molecular orbitals, which are in close energetic proximity to the Fermi level. In such a case, the current becomes only weakly distance-dependent with a low β value. Reversible peak-shaped I–V characteristics indicated that part of the conduction mechanism involves resonant tunneling. At first glance, the current decreased with increasing distance between tip and substrate. Nevertheless, the β-values for the oPE molecules remained low (i.e. around 0.15 Å^{-1}). Upon closer inspection, a strong tip bias dependence was noted. A possible rationale involves that molecular orbitals are affected by interactions with the contacts and by the applied voltage.

Spectroscopic methods, also, provide a direct means of addressing the issue of energy matching. For example, one-photon photoelectron spectroscopy provides access to the energies and characteristics of the occupied electronic levels. Two-photon photoelectron spectroscopy allows, on the other hand, the investigation of unoccupied levels [59, 60]. Another example is spectrophotoelectrochemistry. It is possible to directly address the occupied and unoccupied energy levels and characterize them spectroscopically to gain their relative positions.

Finally, we should mention ab initio quantum chemical methods. They yield almost exact values for the energy levels that are involved in the transport processes. They also assist in supporting the experimental data used for assessing the transport properties of a system. However, these calculations are often limited by the dimensions of these systems and can only be applied to relatively small structures.

4.3 Specific Aspects of Photoinduced Electron Transfer in Organic π-Conjugated Systems

In previous chapters, detailed theoretical principles of CT processes in general and photoinduced CT processes in particular have been provided. The next step is the practical application of these principles to systems which will be discussed within the scope of this thesis, namely organic π-conjugated donor–acceptor supramolecular assemblies. Furthermore, we will draw our attention to CT processes, which are triggered by photochemical stimuli. They provide the theoretical background, which is required to understand the charge-transport properties of wire-type molecular bridges.

4.3.1 Background

Photoinduced charge-transfer reactions have been extensively investigated in many fields of science for more than 15 years. Especially, in the now fast-developing field of photovoltaics it is of fundamental interest to understand the photophysics and photochemistry of excited states in organic molecules [61]. Photosynthetic energy conversion in green plants serves herein as the ultimate prototype [62].

As already outlined in the previous parts, the basic description of photoinduced charge transfer between a donor D and an acceptor A considers different steps. For the following discussion we will assume the following reaction sequence:

- Step 1: $D + A \rightarrow D^\star + A$ (photoexcitation of D).
- Step 2: $D^\star + A \rightarrow (D-A)^\star$ (excitation delocalized between D and A).
- Step 3: $(D-A)^\star \rightarrow (D^{\delta+} - A^{\delta-})^\star$ (polarization of excitation: partial charge transfer).
- Step 4: $(D^{\delta+} - A^{\delta-})^\star \rightarrow (D^{\bullet+} - A^{\bullet-})$ (formation of a radical ion pair).
- Step 5: $(D^{\bullet+} - A^{\bullet-}) \rightarrow D^{\bullet+} + A^{\bullet-}$ (complete charge separation).

Alternatively, we can formulate a hole transfer from an excited acceptor onto a donor. In general, A and D should be either covalently linked (intramolecular CT)

4.3 Specific Aspects of Photoinduced Electron Transfer

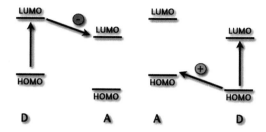

Fig. 4.1 Energy-level alignment for electron transfer from D to A (*left*) and for hole transfer from D to A (*right*)

or spatially very close (intermolecular CT). At this point, we will focus on the intramolecular CT, because the investigated systems contain molecular bridges, however, which link donors and acceptors.

Importantly, to initiate step 2, the electronic wavefunctions of A and D require a significant coupling. In our case this is given by the molecular bridge. On the other hand, a relatively small distance between the donor and acceptor sites would also meet this requirement [63, 64].

Step 4, otherwise, requires an alignment of the energy levels of the participating components. Only if the offset between the LUMOs (for electron transfer) and the HOMOs (for hole transfer) of donor and acceptor is large enough to overcome coulombic attractions between charges, charge separation may occur. The energetic difference between initial and final state represents the driving force for the CT reaction. This is schematically depicted in Fig. 4.1. When donor and acceptor are covalently linked, the CT reaction is then mediated by the transport properties of the intervening linker. In general, the energetics and the kinetics of a photoinduced charge separation are described by the theories of Marcus and Jortner taking into account the orientation and the distance between the donors and the acceptors [65–67].

In some cases, the charge transfer state might be metastable due to a fast delocalization of charge carriers on one or on both molecules [68].

4.3.2 The Classical Marcus Theory

To fully understand the relaxation pathways for photoinduced charge-transfer reactions in solutions we need to take solvent effects into account. For that reason it is necessary to recall some basic principles of the classical Marcus Theory for electron-transfer reactions in solution.

Electron transfer profoundly affects chemical reactivity by inverting normal electron densities in electron donor/acceptor pairs and therefore activating previously inaccessible reaction modes. The basic principles have been widely discussed in several reviews [69–72].

Employing the Franck–Condon principle, i.e. preservation of the nuclear configuration of reactant and product at the point of transition, we can assume a horizontal transition between the donor (D) and acceptor (A). In terms of the previously introduced potential energy surfaces, both the reactants (DA) and products (D^+A^-) posses multidimensional potential energy surfaces, which are functions of many nuclear inter- and intramolecular coordinates. Several coordinates, including the separation distance of the reactants, the molecular geometries of the reactants and the orientation of solvent molecules, undergo significant changes during the course of the reaction. For that reason, in transition state theory, a reaction coordinate is introduced to account for these changes and to reduce the multidimensional potential energy surface to a one-dimensional reaction profile.

Marcus developed a theory [73], which represents the reaction system in Gibbs (free) energy space and approximates the Gibbs energy profiles along the reaction coordinate as parabolas with equal curvatures. As a consequence, the change of entropy in the investigated systems, which usually undergo extremely fast reactions, is assumed to be nearly zero. Then, the change of Gibbs energy is equal to the change of enthalpy.

Figure 4.2 illustrates the parabolic free-energy surfaces as a function of reaction coordinate. In particular, three different kinetic regimes are shown in accordance with the classical Marcus theory. The reorganization energy, λ, represents the change in free energy upon transformation of the equilibrium conformation of the reactants to the equilibrium conformation of the products when no electron is

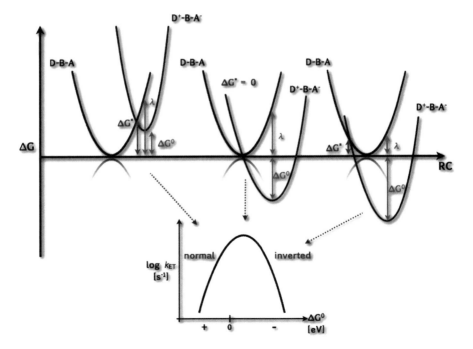

Fig. 4.2 The free energy Marcus regimes (*top*) and the corresponding dependence of the ET transfer rates on the free reaction energy (*bottom*). RC is the reaction coordinate

transferred. λ is the sum of the inner (λ_i) and outer (λ_o) reorganization energies, which account for the changes of structure—such as bond distances and angles—and for the changes of solvation, respectively. The free reaction energy, $-\Delta G^0$, is the difference in free energy between the equilibrium conformations of the reactants and the equilibrium conformation of the products. Thus, $-\Delta G^0$ represents the driving force of the reaction. Dependent on the relationship between λ and $-\Delta G^0$, we can derive an expression for the free activation energy for electron transfer, which corresponds to $\Delta G^\star$:

$$\Delta G^\star = \frac{\lambda}{4}\left(1 + \frac{\Delta G^0}{\lambda}\right)^2. \tag{4.6}$$

With Eq. 4.6 we can formulate the free-activation-energy dependence of the rate constant for electron-transfer reactions

$$k_{ET} = A \exp^{\frac{\Delta G^\star}{k_B T}} \tag{4.7}$$

with the Boltzmann constant k_B and the temperature T.

The exponential term of 4.7 in conjunction with 4.6 contain an important prediction, namely that three distinct kinetic regimes exist, depending on the driving force of the electron transfer process. The three kinetic regimes are also shown schematically in Fig. 4.2 (lower part) in terms of the classical Marcus parabolas:

1. The *normal regime* ($\Delta G < \lambda$ and $-\Delta G < \lambda$) for small driving forces, where the ET rate increases with decreasing activation energy $\Delta G^\star$. The activation energy $\Delta G^\star$ decreases with increasing $-\Delta G^0$.
2. The *activationless regime* ($\Delta G \approx -\lambda$) where the change of the driving force ΔG does not cause large changes in the reaction rates. In this regime, no activation energy is required, i.e. $\Delta G^\star \approx 0$.
3. The *inverted regime* ($-\Delta G > \lambda$) for strongly exergonic processes leads to a decrease of the ET rate.

As an outcome, the most propitious situation that guarantees fast charge-separation rates and slow charge-recombination (back-electron transfer) is when the charge-separation processes occur either in the *normal* or in the *activationless* Marcus regimes, whereas the charge-recombination is located in the *inverted* region of the Marcus parabola. In this scenario the rate of the strongly exergonic charge-recombination process will be automatically minimized.

Determining the feasibility of an electron-transfer pathway for a bimolecular ET reaction usually involves the free reaction energy change $(-\Delta G^0)$ that is calculated via the Weller equation [74]:

$$\Delta G = -E_{00} + E_{D^+/D^0} - E_{A^0/A^-} - E_{IP}, \tag{4.8}$$

where E_{00} is the energetic difference between the S_0 and the S_1 states, E_{D^+/D^0} is the oxidation potential of the electron donor, E_{A^0/A^-} is the reduction potential of the electron acceptor and E_{IP} is the ion pair stabilization energy

$$E_{IP} = \frac{1}{4\pi\varepsilon_0} \frac{e^2}{\varepsilon_S d_{IP}} \qquad (4.9)$$

with e being the electron charge, ε_S the dielectric constant of the solvent and d_{IP} the distance between the donor and acceptor moiety. The free energy, as given by Eq. 4.8, is often referred to as the energy gap.

4.3.3 Photoexcitation and Relaxation Processes in Solution

Photoexcitation as the induction mechanism for CT reactions has been studied extensively and the undergoing processes are nowadays well understood [75]. It is also a "cheap" mechanism for generating free charge-carriers, which in turn can be used elsewhere.

To fully understand the photoexcitation processes we should ask ourselves how light is absorbed by a molecular species. Understanding this phenomenon gives insight into the parameters that trigger CT processes in molecules. As we have elucidated in the previous sections, the transfer of electrons from one end of a large molecule to another is based on electronic interactions between the molecular subunits. In fact, this interaction may be induced by the absorption of light.

4.3.3.1 Photoabsorption

Any form of charge transfer processes requires external stimuli and probes. In general, these stimuli are in the form of wave-like excitations (such as electromagnetic or pressure waves) or of chemical nature (such as electrons, protons, molecules or ions). In this respect, photoabsorption or photoexcitation is perhaps the most general way to couple a molecular system to its environment. Many chemically driven processes such as electron transfer, energy transduction, optoelectronics, photodetection, and optical switching depend all on the absorption of light. In other words, these processes require energy absorbed in form of electromagnetic waves such as sunlight.

The nature of the molecular system implements a change in the physical mechanism of the photoabsorption process. Once again we may, however, employ the golden rule expression. We use it in a general sense to express the absorption rate in the form of:

$$W_{g \to x} = \frac{2\pi}{\hbar} (\langle x | \mu | g \rangle)^2 \rho_f. \qquad (4.10)$$

4.3 Specific Aspects of Photoinduced Electron Transfer

For a direct absorption, x and g label the excited and ground states, respectively. These states are, in turn, coupled by a dipole moment operator μ, and are assumed to be of Born–Oppenheimer type, i.e. the electronic contributions are separated from those of the nucleus. With these assumptions, the interactions between these states result from vibrational overlaps between the ground and excited state with the transition probability $W_{g \to x}$ and the density of the final states ρ_f. In general terms, we now can evaluate the transition probability, which mainly depends upon two parameters:

1. the strength of the coupling between the initial and the final states,
2. the number of transition pathways, i.e. density of the final states.

As a matter of fact, the transition probability relates to the mean lifetime τ of the state by $W_{g \to x} = \frac{1}{\tau}$. Now, we have developed an expression for the decay probability. Obviously, the transition rate will increase if the coupling between the initial and final states, represented by the matrix element $\langle x| \mu |g \rangle$, is strong. The ground and excited states typically differ from each other energetically. The two electronic states that we consider are represented by two energy levels. Figure 4.3 (left) sketches this for a simple one-site, two-level model. Here, the photoabsorption process is simply a transition from the ground state to the excited state followed by a deactivation process. A more complex situation is present, when we deal with two-site situations, such as donor/acceptor, or or even multiple-site, donor–bridge–acceptor systems. The interaction between the different sites leads to a mixing of the different levels, as demonstrated in Fig. 4.3 (right). Then, the spectroscopic states evolve from the mixing of the zero-order states. These consist, for example, of the neutral donor/acceptor system (DA), the photoexcited donor/acceptor system $((DA)^\star)$, and the charge-separated donor/acceptor system (D^+A^-). Implicit is that photoabsorption and subsequent radiative decays are substantially altered relative to the two-level model given in Fig. 4.3 (left) [76].

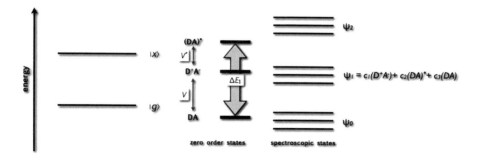

Fig. 4.3 A simplified two-level model for photoabsorption (*left*). The two levels are represented by the corresponding wavefunctions (g, ground state; x, excited state) with the matrix element $\langle x | \mu |g \rangle$ giving the transition probability between the ground and excited state. In the *right* part a simple three-state model for a two-site situation represents the mixing between the zero order states (neutral, excited and charge-transfer) to form the spectroscopic states

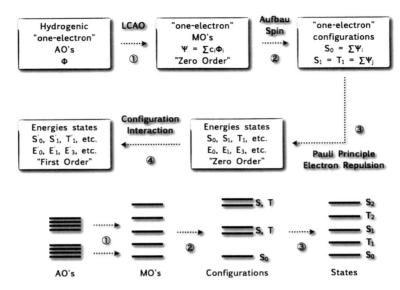

Fig. 4.4 Schematic of the method of building an energy diagram from theory. Each of the final states in the "zero order" is derived from a distinct configuration. The higher the level of approximation (e.g. configuration interaction), the more configurations are required to describe each state

For the sake of simplicity, we will try to illustrate the method of building up the spectroscopic states from atomic orbitals (AOs) for a "one electron" situation. Figure 4.4 summarizes the pathway from "one electron" atomic orbitals (ϕ), to "one electron" molecular orbitals (ψ), to one electron configurations $\Pi_i \psi_i$, to Zero Order States, to First Order States, and finally to First Order or "working" states and state energy diagrams. Interestingly, the working state is at the level of a "two orbital configuration approximation", i.e. we take only two orbitals into account, and characterize a state in terms of the orbital configuration of the two highest-energy electrons [77].

In principle, a molecule in a particular electronic state may exist in various configurations of its nuclei. Hereby, each configuration in space corresponds to a particular potential energy of the system. A map of the potential energy versus the nuclear configuration for a given electronic state is called the *potential energy surface*. However, a two-dimensional energy curve is more readily visualized than a three-dimensional energy surface. Thus, we shall show how the simple notion of *potential energy curves* for a diatomic molecule may be used to unify the ideas of structure, energetics and dynamics and then extrapolate the concepts to general energy surfaces [78, 79].

In particular, radiative and non-radiative transitions between states and photochemical reactions may be visualized under the framework of energy surfaces. Often, reactions cannot be interpreted in terms of the motion of the system along only one energy surface but rather must be characterized by transitions between interacting potential energy surfaces. For instance, a nonadiabatic reaction is a reaction in which a transition from one electronic potential energy surface to another occurs during the reaction. All photoreactions which lead to stable ground state molecules are nonadiabatic reactions, since light absorption automatically places the system onto an excited state surface. However, the resulting products are usually observed on the ground state surface. This leads to the assumption that a transition between two electronic surfaces must have occurred at some point [80, 81].

When dealing with photochemical systems, two broad distinct classes of nonadiabatic transitions have to be discussed:

1. Radiative nonadiabatic transitions, e.g. fluorescence and phosphorescence.
2. Radiationless nonadiabatic transitions, e.g. internal conversion, intersystem crossing, and most photoreactions.

Nonadiabatic transitions must obey the Franck–Condon principle, such that the initial and final geometries must be very similar. As a result, a "chemical reaction", i.e. a reaction involving drastic changes in nuclear geometry, never occurs via a radiative nonadiabatic transition. However, many photochemical reactions employ radiationless nonadiabatic transitions as a key step, which affects either the reactivity and/or the efficiency of the reaction [82–84].

Understanding the theoretical principles of light induced nonadiabatic reactions is therefore crucial for the comprehension of the photo-driven processes that lead to photoinduced charge transfer and energy transfer reactions which will be discussed later on in this thesis.

4.3.3.2 The Franck–Condon Principle and Radiative Transitions

The classical model of the Franck–Condon principle predicts the electronic transition occurring from the *most probable* nuclear configuration of the ground state—which is the static, minimum equilibrium arrangement of the nuclei—to the excited state surface. Dependent on the position of the ground state and excited state surfaces relatively to each other, the transition would occur to a certain vibrational level (v). At the completion of such electronic transition the nuclei are static, but in a new force field of the new electronic distribution. As a consequence, they begin to vibrate away from and back to their initial arrangement. It follows that the original arrangement is a turning point of the new motion, and the vibrational energy (ΔE_{vib}) is stored by the molecule. The velocity of this motion depends on the excess of kinetic energy. Figure 4.5 represents a schematic representation of the classical model.

In terms of quantum mechanics, a system with "zero" energy is impossible. A quantum system must possess a minimum energy of E_v. This postulate is due to the irrepressible zero-point motion imposed on microscopic systems by the uncertainty principle and by quantization. Thus, the classical concept of nuclei in space and associated motion is replaced by the concept of a nuclear or *vibrational*

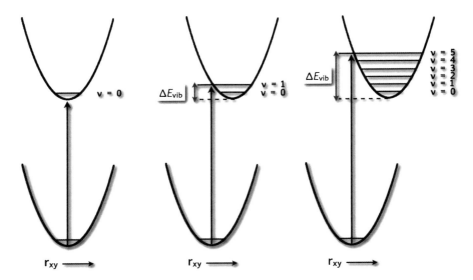

Fig. 4.5 Schematic representation of the Franck–Condon principle for radiative transitions

wavefunction, χ, which "codes" the nuclear configuration and momentum but is not as restrictive in confining the nuclear configurations to the regions of space bound by the classical potential energy curves. In classical mechanics we consider that electronic transitions require similar nuclear configurations and momentum in the initial and final states at the instant of transition. In quantum mechanics this requirement becomes the *net positive overlap* of the wavefunctions in the initial and final states during the transition. This overlap is given by the Franck–Condon integral $\langle \chi_i \mid \chi_f \rangle$. The probability of any electronic transition is directly related to the square of the vibrational overlap integral, i.e. $\langle \chi_i \mid \chi_f \rangle^2$, which is called the *Franck–Condon factor*.

The Franck–Condon factor governs the relative intensities of vibrational bands in the electronic absorption and emission spectra. In radiationless transitions the Franck–Condon factor is also important to determine the transition rates. Since the value of $\langle \chi_i \mid \chi_f \rangle^2$ tracks that of $\langle \chi_i \mid \chi_f \rangle$, we will generally consider the integral itself rather than its square. The larger the difference in the vibrational quantum numbers *i* and *f*, the more the molecular geometries and momentum of the initial and final states will differ from each other, and the more difficult is the transition. Indeed, this is exactly the result what one might expect from the Franck–Condon principle. In other words, the product $\langle \chi_i \mid \chi_f \rangle$. is related to the probability that a state ψ_i will or will not have the same shape and momentum as ψ_f. Thus, the Franck–Condon overlap integral is analogous to the electronic overlap integral $\langle \psi_i \mid \psi_f \rangle$. In consequence, poor overlap means weak interactions and slow transition rates.

Figure 4.6 schematically represents the quantum mechanical basis of the Franck–Condon principle for radiative transitions. As a result, the process of

4.3 Specific Aspects of Photoinduced Electron Transfer

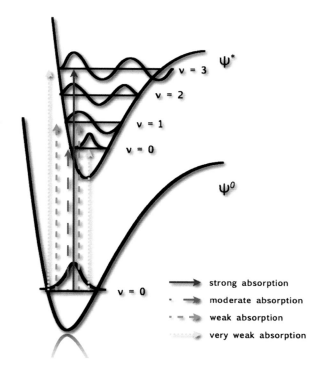

Fig. 4.6 In the quantum mechanical version of the Franck–Condon principle, the molecule undergoes a transition to the upper vibrational state that most closely resembles the vibrational wavefunction of the vibrational ground state of the lower electronic state. The wavefunctions with the greatest overlap integral of all the vibrational states are responsible for the strongest absorption

photoabsorption is assumed to initiate form the $v = 0$ level of ψ^0. The most likely radiative transition from $v = 0$ of ψ^0 to a vibrational level of $\psi^\star$ will correspond to the transition for which χ_0 and $\chi_i^\star$ is maximal. In our representation, this corresponds to the $v = 0$ to $v = 3$ transition. Other transitions from $v = 0$ of ψ^0 to vibrational levels $v \neq 3$ of $\psi^\star$ will occur but with lower probability.

The same general ideas will apply to emission, except the important overlap is then between χ_0 of $\psi^\star$ and the various vibrational levels of ψ^0 [85].

4.3.3.3 The Franck–Condon Principle and Radiationless Transitions

Generally, the occurrence of unimolecular radiationless transitions such as internal conversion and intersystem crossing may be inferred from quantum yield measurements. The common experimental observation in such cases is the lack of a net reaction after absorption of a photon. The Franck–Condon principle that implies radiative transitions with quantum yields of less than unity also applies to radiationless processes, as it prohibits vertical transitions between surfaces separated by large energy gaps and favors those at Zero Order surface crossings.

Suppose that a molecule starts off on an excited surface $\psi^\star$ and makes a trajectory from A to B on the $\psi^\star$ surface during its zero-point motion. Classically,

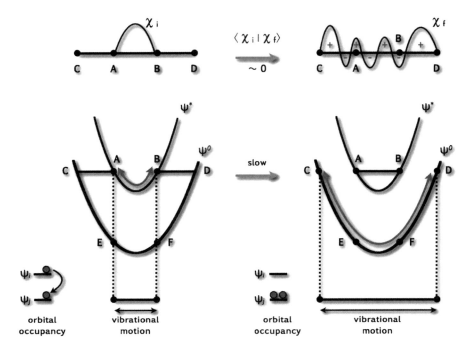

Fig. 4.7 Visualization of the quantum mechanical basis for a slow rate of radiationless transitions

a transition to the energetically lower ψ^0 surface will require an abrupt change in geometry or an abrupt change in kinetic energy. The net result of either transition is that the vibration of the molecule will abruptly change from a placid, low-energy vibration between points A and B to a violent, high-energy vibration between two new points C and D. Now, both the positional and momentum characteristics will change drastically. Electrons resist, however, to drastic changes in orbital motions and spatial locations, whereas nuclei resist to drastic changes in their vibrational motions of spatial geometries. Not surprisingly, the wavefunction χ_i is completely different in form (i.e. positive everywhere with no node) from that of χ_f (i.e. positive and negative with several nodes). This situation is schematically visualized in Fig. 4.7.

For vertical transitions from $\psi^\star$ to ψ^0, for example, the rate-limiting step is the electronic perturbation, which occurs first. Nuclear motion is now suddenly controlled by the ψ^0 surface rather than by $\psi^\star$. In fact, vibrations along ψ^0 must compensate the extinct energy that is associated with the transition from $\psi^\star$ to ψ^0. The horizontal transition, on the other hand, would be given by the rate-limiting nuclear geometry perturbation. After that the promotion from one geometry to another takes place. The electronic motion suddenly switches from that of $\psi^\star$ to that of ψ^0. In this scenario, vibrations which bring $\psi^\star$ from one geometry to the another may also act as mediator to dissipate the excess energy.

4.3 Specific Aspects of Photoinduced Electron Transfer

It is important to distinguish between two general situations, namely crossing and not-crossing of the ψ^0 and the $\psi^\star$ surfaces. The poor overlap of vibrational wavefunctions of ψ^0 and $\psi^\star$ for a molecule in the lowest vibrational level of $\psi^\star$ for the noncrossing situation contrasts with the significant overlap for the crossing situation.

In both cases, χ_i corresponds to the $v = 0$ level of $\psi^\star$ and χ_f corresponds to the $v = n$ level of ψ^0. The extra electronic energy (ΔE_{el}) that converts into vibrational energy and the vibrational quantum number (v) of the state produced by the transition are identical. Nevertheless, the radiationless transitions with a good overlap between χ_i and χ_f—high $\int \chi_i \chi_f$—will evidently occur much faster than the radiationless with poor overlap—low $\int \chi_i \chi_f$. As a consequence, we postulate that the radiationless transition, where no surface crossing is present, is Franck–Condon forbidden, because $\langle \chi_i | \chi_f \rangle$ is nearly 0. In contrast, a good overlap $\int \chi_i \chi_f$ leads to Franck–Condon allowed transitions with $\langle \chi_i | \chi_f \rangle > 0$.

In conclusion, we can say, that with respect to the Franck–Condon factors, a radiationless transition from $\psi^\star$ to ψ^0 will be very slow when the net positive overlap is poor, i.e. $\langle \chi_i | \chi_f \rangle$ is nearly 0 and fast with $\langle \chi_i | \chi_f \rangle > 0$ [85, 86].

In general, the electronic absorption and emission spectra provide important information about the structure, energetics and dynamics of electronically excited states. With the theoretical background in hands, we may be able to predict the deactivation processes of lightinduced excitation, such as radiative and radiationless transitions, energy transfer and intersystem crossing processes. From spectroscopic measurements we can deduce the $\psi^0 + h\nu \rightarrow \psi^\star$ absorption processes, as well as the $\psi^\star \rightarrow \psi^0 + h\nu$ emission processes, which provide us a fairly complete "static" state energy diagram. From lifetime measurements and emission efficiency measurements, dynamics of the photophysical and photochemical pathways available to $\psi^\star$ can be determined.

4.3.3.4 Relaxation Processes Following Photoexcitation

In general, all photoexcitation processes require a broadly absorbing chromophore, which should have large oscillator strengths and relatively slow radiative decays. In DBA systems, the coupling of the chromophore to the terminal sites of the system depends on tuning the intramolecular relaxation channels from the initial to the final state—excited and charge-separated states, respectively. Taking the aforementioned in context, the design of a molecular system capable of electron transfer reactions needs to fulfill the following requirements:

1. Implementation of chromophoric units that absorb light in the desired range of the solar spectrum.
2. Significant electronic coupling V between donor and bridge as well as between bridge and acceptor, represented by the matrix element $\langle x | \mu | g \rangle$, must exist. In other words, the transition between the initial (neutral), intermediate (locally excited) and final (charge separated) state must be Franck–Condon allowed ($\langle \chi_i | \chi_f \rangle > 0$).

The coupling is responsible for mixing of spectroscopic states. These, in turn, are the intramolecular relaxation channels, such as energy gradients to guide electrons unidirectionally. High energy gradients and therefore straightforward pathways for electrons to proceed enables a high control of charge-transfer reactions and leads to kinetically stable photoproducts. Simply speaking, an excited molecule must be sufficiently stable and sufficiently energetic to react, i.e. transfer the electron, at a rate competitive to other pathways of excited-state deactivation [87].

The Arrhenius expression

$$k = A \exp^{-\frac{E_a}{RT}} \quad (4.11)$$

correlates the reaction rate with the activation energy for a specific reaction and a preexponential factor A. The previous chapters have shown that the preexponential factor is composed of several structural and electronic properties, e.g. electronic coupling, Franck–Condon factors, etc. Needless to say that, determining the electron transfer rate can easily turn into a very sophisticated issue and needs an accurate understanding of not only the electron transfer processes itself but also the designed molecular structure.

Importantly, all photoinduced processes share some common features. A photochemical reaction starts with the ground state structure, proceeds to an excited state structure and ends in the ground state structure. Thus, photochemical mechanisms are inherently multistep and involve intermediates between reactants and products. In the course of a photoinduced charge transfer reaction the molecule passes through several energy states with different activation barriers. This renders the electron transfer pathway quite complex.

However, if the nature (e.g. electronic configuration) of the first singlet excited state S_1 and the first triplet excited state T_1 are known, the electron-transfer is easy to handle in a mechanistic sense. An important preequisite is to know if the S_1 and/or T_1 are the exclusive precursor states for the electron-transfer product or if intermediates such as upper excited states, biradicals or ground-state intermediates might be involved. In addition, structural properties have to be considered as well because the states of a chemical system are parametrized in terms of distinct structures. For instance, not all structural interconversions are allowed under a given set of reactions conditions (e.g. solvent or temperature) [88].

Next, we should survey the fate of the excited states and the different relaxation mechanisms. Such processes may follow a photoinduced excitation of organic chromophores in solution. An overview of the different relaxation pathways in solution will be presented in the following sections. In addition, we will discern practical ways to identify these processes by conventional spectroscopic methods.

In general, we find four major categories for the various relaxation processes in solutions. These range from electronic relaxation and solvent relaxation to orientational relaxation, and finally vibrational relaxation [89]. Stationary and time-resolved spectroscopy provides powerful means to explore the electronic and solvent relaxations. Nevertheless, the many different and extremely fast processes involved in the excited-state deactivation make it difficult to gather a pellucid picture of the molecular dynamics taking only the reactants and the products into account.

4.3.3.5 Characterization by Stationary Spectroscopy

In condensed media and liquid phase, the excess vibrational energy is typically released to the surrounding medium on a very fast timescale and therefore not observable by steady-state methods. Following the vibrational relaxation, the molecule returns to its electronic ground state. This occurs upon emission of a photon or via radiationless transitions. Radiationless transitions should be divided into two general classes [90]:

1. Internal conversion, where the spin-state of the molecule is preserved (singlet → singlet or triplet → triplet).
2. Intersystem crossing, where the spin of the electron changes (singlet ↔ triplet).

The emission behavior is embodied in two rules: Kasha's rule and Vavilov's law [91]. Kasha's rule states that if a molecule emits a photon, it will always originate from the lowest excited state of a given spin multiplicity. If the emission occurs from the lowest singlet excited state it is called *fluorescence* and when the lowest excited state is a triplet it is known as *phosphorescence*. In addition, Vavilov's law implies that the fluorescence quantum yield (Φ_F) is essentially independent on the excitation wavelength.

The fluorescence quantum yield, Φ_F, is defined as the ratio between the number of photons emitted and the number of photons absorbed. Using the rates of the relaxation pathways it can be expressed via:

$$\Phi_F = \frac{N_{emit}}{N_{abs}} = \frac{k_{rad}}{k_{all}} = \frac{k_{rad}}{k_{rad} + k_{IC} + k_{ISC} + k_{ET}} = \frac{1/\tau_{rad}}{1/\tau_{fl}}. \quad (4.12)$$

Here, k_{rad} is the rate of radiative relaxation to the ground state, k_{IC} is the rate of internal conversion, k_{ISC} is the rate of intersystem crossing, and k_{ET} is the rate of the energy/electron transfer. τ_{fl} is the *fluorescence lifetime* [92]. Assuming that spontaneous emission is the only possible deactivation mechanism, the average time that the molecule exists in the excited state is referred to the radiative lifetime τ_{rad}. τ_{rad} depends on the corresponding electronic transition and to some extent on the surrounding environment [93]. Importantly, the solvent might influence τ_{rad} either by its refractive index or directly by affecting the transition moment [92].

4.3.3.6 Characterization by Time-Resolved Spectroscopy

To resolve excited state processes in terms of kinetics and spectroscopy the course of a reaction needs to be followed as a function of time. Charge transfer processes typically occur on picosecond to nanosecond time scales. To this end, ultra-fast laser spectroscopy, e.g. femtosecond transient absorption measurements and nanosecond flash photolysis, evolved as important techniques to gather kinetic and spectroscopic information on the observed phenomena. In general, pump–probe techniques are utilized within these spectroscopic methods. By spatially crossing two laser beams in a sample, one excites the molecule (pump) and the other is used to detect the spectral response of the sample (probe). The herein applied techniques will be presented in the experimental section in more detail (Chap. 7).

4.3.3.7 Internal Conversion

In competition to electron transfer processes is internal conversion (IC), in which deactivations of excited states occur via a nonradiative transition to the electronic ground state. Visualizing IC rates is practically impossible because of the lack of a direct probe mechanism for nonradiative transitions. A notable approach to overcome this lack of detectability implies fluorescence quantum yield measurements, which is, however, only indirect.

Regarding the fact, that IC is a radiationless transition between isoenergetic levels of different electronic states with equal multiplicities, the rates must be comparable to or faster than vibrational relaxation. In fact, the rates depend on the energy gap between the lowest vibrational levels of the electronic states involved. It has been shown experimentally that for aromatic hydrocarbons the radiationless transition from S_1 to S_0 is negligible if the energy difference between these two electronic states, $E_{S_1-S_0}$ exceeds 60 kcal/mol. Thus, for molecules with low-lying singlet states, such as highly conjugated π-systems, relaxation of the excited state by IC plays an increasing role and may account for more than 90% of the S_1 state deactivation [91].

Summarizing, we can express the rate of internal conversion by the following exponential relationship:

$$k_{IC} = 10^{13} \exp^{-\alpha \Delta E_{S_1-S_0}} . \tag{4.13}$$

This equation, which is also referred to as the energy-gap law, expresses the dependence of the rate constant k_{IC} on the energy gap $\Delta E_{S_1-S_0}$. The proportionality constant α is characteristic for the investigated molecular system. For aromatic systems containing phenyl rings this value is 4.85 eV^{-1} [92].

4.3.4 Influence of the Solvation on the Electronic Relaxation Dynamics

Within the scope of this work, all experimental results have been generated in solvent environments. For that reason, it is important to rationalize the influence of the solvent on the investigated processes.

4.3 Specific Aspects of Photoinduced Electron Transfer

In general, solvation refers to the surrounding of a solute (i.e., molecule or ion) by a shell of more or less tightly bound solvent molecules. The solvent shell results from intermolecular forces between solute and solvent. These forces depend on the polarity of the solvent and can be related to the solvation energy. The latter is considered as the change in Gibbs free energy upon transferring an ion or molecule from vacuum (or gas phase) into a solvent environment. Often the term Gibbs energy of solvation is used, which describes the ability of solvation for a particular solvent. The solvation ability depends on four main components [94]:

1. The *cavitation energy*—energy required to generate a "hole" in the solvent matrix to dissolve the solute.
2. The *orientation energy*—partial reorientation of solvent molecules caused by the presence of the solute.
3. The *isotropic interaction energy*—long-range forces experienced by the solute, e.g. electrostatic forces, polarization and dispersion energy.
4. The *anisotropic interaction energy*—specific formation of hydrogen bonds or donor–acceptor electron pair bonds at well localized areas of the solute.

In summary, solvent interactions play a very important role that control photoinduced electron-transfer processes. It is mainly the high impact on the stabilization energies of the generated high-energy species that are decisive. We have to distinguish between two different effects—the static solvent influence and the dynamic solvent influence.

4.3.4.1 Static Solvent Influence

Solvents alter the energies of the reactants and products. In particular, the potential energy surface, on which the reaction occurs, changes when going from gas-phase to solvent environment. This occurs in a static sense and generates a significant energy difference between the ionization potentials. For instance, the ionization potential in a condensed liquid phase (I_{liq}) and in the gas phase (I_{gas}), differs via [77, 96]:

$$I_{liq} = I_{gas} + P_+ + V_0. \tag{4.14}$$

P_+ is the adiabatic electronic polarization energy of the medium, experienced by a positive ion. V_0 is the minimum energy of a "quasi-free" electron in the liquid relative to an electron in vacuum. Then, if we consider the solvent as a dielectric continuum, Born's equation [97] provides an expression to estimate P_+:

$$P_+ = \frac{e^2}{2r_+}\left(1 - \frac{1}{\varepsilon_\infty}\right). \tag{4.15}$$

e is the elementary charge, r_+ is the effective ionic radius and ε_∞ is the optical dielectric constant of the solvent. With the help of these two equations, we can then determine the free energy of solvation for a solute. In turn, it is possible to estimate the solvent effects as they may impact radical ion pairs.

4.3.4.2 Dynamic Solvent Influence

By far more interesting are dynamic solvent effects due to their influencing role on the kinetics of electron transfer processes in general. In fact, in the course of a charge-transfer reaction, the energy and momentum undergo significant changes. Furthermore, the charge distribution will change dramatically and the solvent molecules respond to these rearrangements by exchanging energy and momentum with the reacting species. In this regard, the solvent assumes a more active role. Contrary to static solvent effects, which almost exclusively influence the free activation energy, $\Delta G^{\star}$, dynamic solvent effects are often discussed in terms of friction. The origin of that could be either collisional or dielectric. They both directly affect electron transfer rates, because they appear in the frequency factor A of Eq. 4.7.

Dynamic solvent effects appear to be of fundamental importance for electron and energy transfer reactions. Implicit is, a "dielectric friction", which happens to be particularly striking in solvents with high dielectric constants, i.e. polar solvents. Thus, the friction effect increases with increasing solvent polarity. The existence of a coupling between the solvent and the reacting system, which is usually of electrostatic origin, leads to dynamic solvent effects. If we consider electron transfer reactions, where spatially separated radical ion pairs are formed, stabilization effects by electrostatic forces and polar interactions will be very strong. To this end, it is reasonable to assume that the reaction rates are strongly affected by the polarity of the solvent. The free activation energy $\Delta G^{\star}$ is the factor that is most important [98].

Such solvent effects are usually referred to as *solvatochroism*. It reflects the influence of solvent polarity on excited state energies and charge transfer energies. Typically, this is seen as an additional shift of the emission characteristics. It can be explained by the reorganization of solvent molecules, when considering, for example, the change of dipole moment of the solute in order to stabilize the system in the potential energy minimum. In general, the energies of the electronic states are lowered by solvent interactions. Solvatochroism may be positive (red-shift) or negative (blue-shift) and depend on whether the excited state or the ground state is more effectively stabilized. As a common rule, excited or charge-separated states tend to be more polar than ground-states. Thus, positive solvatochroism is observed for $\pi \rightarrow \pi^{\star}$ or charge-transfer transitions. Since these effects increase when increasing the solvent polarity, spectroscopic methods help to identify and quantify them [92, 94, 99].

Experimentally, these effects are tested by fluorescence and absorption measurements. These directly probe solvent polarization dynamics on molecular timescales [100, 101]. For instance, the time resolved fluorescence spectrum of a chromophore, whose excited state dipole moment is subject to interactions with the surrounding solvent molecules, will exhibit fluorescence spectra that are strongly solvent dependent. The solvent molecules attempt to compensate the changes of charge density in the chromophore and, in sum, the fluorescence behavior will change gradually. In particular, the solvent molecules will reorganize themselves in accordance with the new equilibrium charge distribution. As a consequence, the chromophore emission shifts progressively as it reflects the course of the solvation energy relaxation. In theory, this can be described by the solvation time correlation function:

$$S_{obs}(t) = \frac{v_t - v_\infty}{v_0 - v_\infty}, \quad (4.16)$$

v refers to some characteristic frequencies in the spectrum (e.g. maximum or mean fluorescence frequency) at the times zero and infinity. The spectral response function, $S_{obs}(t)$, is of very complicated nature. Its full theoretical description is far beyond the scope of this thesis but has been discussed in great detail elsewhere [102–103].

It should be noticed that nonpolar solvents exhibit solvation dynamics, which are qualitatively similar to what has been said about polar solvents. Modeling these effects is not trivial, because the dielectric response of solvents cannot be used as an empirical input [104].

References

1. Emberly EG, Kirczenow G (1998) Phys Rev B 58:10911
2. Nitzan A, Ratner MA (2003) Science Washington DC, US 300:1384
3. Davis WB, Svec WA, Ratner MA, Wasielewski MR (1998) Nature 396:60
4. Schatz GC, Ratner MA (2002) Quantum mechanics in chemistry, 2 edn. Dover Publications, Mineola, New York
5. Linderberg J, Ohrn Y (2004) Propagators in quantum chemistry. Wiley, Hoboken, New Jersey
6. Pople JA, Beveridge DL (1970) Approximate molecular orbital theory. McGraw-Hill, New York
7. Jortner J (1976) J Chem Phys 64:4860
8. Ratner MA, Sutin N (1986) Biochim Biophys Acta 811:265
9. Kramers KH (1934) Physica 1:182
10. Anderson PW (1950) Phys Rev Lett 79:350
11. Anderson PW (1959) Phys Rev Lett 115:2
12. Nitzan A (2001) Annu Rev Phys Chem 52:681
13. Berlin YA, Burin AL, Ratner MA (2002) Chem Phys 275:61
14. Grozema FC, Berlin YA, Siebbeles LDA (2000) J Am Chem Soc 122:10903
15. Finklea HO, Hanshewm DD (1992) J Am Chem Soc 114:3173
16. Slowiski K, Chamberlain RV, Miller CJ, Majda M (1997) J Am Chem Soc 119:11910
17. Chidsey CED (1991) Science Washington DC, US 251:919
18. Leland BA, Joran AD, Felker PM, Hopfield JJ, Zewail AH, Dervan PB (1985) J Phys Chem A 89:5571
19. Oevering H, Paddon-Row MN, Heppener M, Oliver AM, Cotsaris E, Verhoeven JW, Hush NS (1987) J Am Chem Soc 109:3258
20. Closs GL, Miller JR (1988) Science Washington DC, US 240:440
21. Klan P, Wagner PJ (1998) J Am Chem Soc 120:2198

22. Osuka A, Maruyama K, Mataga N, Asahi T, Yamazaki I, Tamai N (1990) J Am Chem Soc 112:4958
23. Helms A, Heiler D, McClendon G (1992) J Am Chem Soc 114:6227
24. Weiss EA, Ahrens MJ, Sinks LE, Gusev AV, Ratner MA, Wasielewsi MR (2004) J Am Chem Soc 126:5577
25. Osuka A, Satoshi N, Maruyama K, Mataga N, Asahi T, Yamazaki I, Nishimura Y, Onho T, Nozaki K (1993) J Am Chem Soc 115:4577
26. Barigelletti F, Flamigni L, Balzani V, Collin J-P, Sauvage J-P, Sour A, Constable EC, Cargill AMW (1994) Thompson. J Am Chem Soc 116:7692
27. Creager S, Yu CJ, Bamdad C, O'Connor S, MacLean T, Lam E, Chong Y, Olsen GT, Luo J, Gozin M, Kayyem JF (1999) J Am Chem Soc 121:1059
28. Sachs SB, Dudek SP, Hsung RP, Sita LR, Smalley JF, Newton MD, Feldberg SW, Chidsey CED (1997) J Am Chem Soc 10:563
29. Sykes HD, Smalley JF, Dudek SP, Cook AR, Newton MD, Chidsey CED, Felberg SW (2001) Science Washington DC, US 291:1519
30. Martin N, Giacalone F, Segura JL, Guldi DM (2004) Synth Met 147:57
31. Atienza C, Martín N, Wielopolski M, Haworth N, Clark T, Guldi DM (2006) Chem Commun (Cambridge, UK) 30:3202
32. Benniston AC, Goulle V, Harriman A, Lehn J-M, Marczinke B (1994) J Phys Chem A 98:7798
33. Osuka A, Tanabe N, Kawabata S, Speiser IS (1996) Chem Rev Washington DC, US 96:195
34. Marczinke B (1994) J Phys Chem A 98:7798
35. Osuka A, Tanabe, Kawabata S, Grosshenny IV, Harriman A, Ziessel R (1995) Angew Chem Int Ed 34:1100
36. Grosshenny IV, Harriman A, Ziessel R (1995) Angew Chem Int Ed 34:2705
37. Marcus RA (1987) Chem Phys Lett 133:471
38. Ogrodnik A, Michel-Beyerle ME, Naturforsch Z (1989) A Phys Sci 44a:763
39. Kilsa K, Kajanus J, Macpherson AN, Martensson J, Albinsson B (2001) J Am Chem Soc 123:3069
40. Lukas AS, Bushard PJ, Wasielewski MR (2002) J Phys Chem A 106:2074
41. Marcus RA (1965) J Chem Phys 43:679
42. McConnell HM (1961) J Chem Phys 35:508
43. Closs GL, Piotrowiak P, McInnis JM, Fleming GR (1988) J Am Chem Soc 110:2652
44. Roest MR, Oliver AM, Paddon-Row MN, Verhoeven JW (1997) J Phys Chem A 101:4867
45. Paddon-Row MN, Oliver AM, Warman JM, Smit KJ, Haas MP, Oevering H, Verhoeven JW (1988) J Phys Chem A 92:6958
46. Jortner J, Bixon M, Langenbacher T, Michel-Beyerle ME (1998) Proc Natl Acad Sci USA 95:759
47. Bixon M, Giese B, Langenbacher T, Michel-Beyerle ME, Jortner J (1999) Proc Natl Acad Sci USA 96:11713
48. Davis WB, Wasielewski MR, Mujica V, Nitzan A (1997) J Phys Chem A 101:6158
49. Kharkats YI, Ulstrup J (1991) Chem Phys Lett 182:81
50. Sourtis SS, Mukamel S (1995) Chem Phys 197:367
51. Felts AK, Pollard WT, Friesner RA (1995) J Phys Chem A 99:2929
52. Cave RJ, Newton MD (1996) Chem Phys Lett 249:15
53. Creutz C, Newton MD, Sutin N (1994) J Photochem Photobiol A 82:47
54. Kumar K, Kurnikov IV, Beratan DN, Waldeck DH, Zimmt MB (1998) J Phys Chem A 102:5529
55. Stuchebrukov AA, Marcus RA (1995) J Phys Chem A 99:7581
56. Golub GH, van Loan CF (1989) Matrix computations. Johns Hopkins University Press, Baltimore
57. Flannery BP, Teukolsky SA, Vetterlink WT (1988) Numerical recipes. Cambridge University Press, Cambridge, UK

References

58. Fan F, Yang J, Cai L, Price Jr DW, Dirk SM, Kosynkin DV, Yao Y, Rawlett AM, Tour JM, Bard AJ (2002) J Am Chem Soc 124:5550
59. Zangmeister CD, Robey SW, van Zee RD, Yao Y, Tour JM (2004) J Phys Chem B 108:16187
60. Zhu XY (2004) J Phys Chem B 108:8778
61. Parts a and d (1988) In: Fox MA, Chanon M (eds) Photoinduced Electron Transfer. Elsevier, Amsterdam
62. Hu X, Schulten K (1997) How nature harvests sunlight. Phys Today 50:28 and references therein
63. Rice MJ, Gartstein YN (1996) Theory of photoinduced charge transfer in a molecularly doped conjugated polymer. Phys Rev B 53:10764
64. Wu MW, Conwell EM (1998) Theory of photoinduced charge transfer in weakly coupled donor-acceptor conjugated polymers: application to an meh-ppv:cn-ppv pair. Chem Phys 227:11
65. Paddon-Row MN (1994) Acc Chem Res 27:18
66. Guldi DM (2002) Chem Soc Rev 31:22
67. Bixon M, Jortner J (1999) Adv Chem Phys 106:35
68. Müller GM, Lupton JM, Feldmann J, Lemmer U, Scharber MC, Sariciftci NS, Brabec CJ, Scherf U (2005) Phys Rev B 72:195208
69. Marcus RA (1964) Annu Rev Phys Chem 15:155
70. Kavarnos GJ, Turro NJ (1986) Chem Rev Washington DC, US 86:401
71. Marcus RA (1993) Angew Chem Int Ed 105:1161
72. Kuznetsov AM, Ulstrup J (1999) Electron transfer in chemistry and biology: an introduction to the theory. JohnWiley and Sons Ltd, New York
73. Marcus RA (1956) J Chem Phys 24:979
74. Rehm D, Weller A (1970) Isr J Chem 8:259
75. Balzani V, Scandola F (1991) Supramolecular Photochemistry. Horwood, Chichester
76. Jortner J, Ratner M (eds.) (1997) Molecular Electronics. Blackwell, London
77. McGlynn SP, Smith FJ, Cilento G. Photochem J (1964) Photobiol A 3:269
78. Dewar MJS, Doughtery RC (1975) The PMO Theory of Organic Chemistry. Plenum, New York
79. Halliday D, Resnick R (1967) Physics. JohnWiley, New York
80. Dauben W, Salem L, Turro NJ (1975) Acc Chem Res 8:41
81. Michl J (1974) Top Curr Chem 46:1
82. Michl J (1972) Mol Photochem 4:243
83. Devaquet A (1975) Top Curr Chem 54:1
84. Dauben W, Salem L (1975) J Am Chem Soc 97:479
85. Atkins P (1974) Quanta: A Handbook of Concepts. Clarendon Press, Oxford
86. Förster T (1970) Pure Appl Chem 24:443
87. Bixon M, Jortner J, Verhoeven JW (1994) J Am Chem Soc 116:7349
88. Turro NJ (1978) Modern molecular photochemistry. Benjamin Cummings, Menlo Park
89. Fleming GR (1986) Chemical Applications of Ultrafast Spectroscopy. Oxford University Press, New York
90. Kasha M (1995) Faraday Discuss 9:14
91. Turro NJ (1991) Modern Molecular Photochemistry 2nd edition. University Science Books, Sausalito, California
92. Klessinger M, Michl J (1995) Excited states and photochemistry of organic molecules. VCH Publishers, Inc., New York
93. Strickler JS, Berg RA (1962) J Chem Phys 37:814
94. Reichardt C (1990) Solvents and Solvent Effects in Organic Chemistry. VCH VerlagsgesellschaftmbH, Weinheim
95. Raz B, Jortner J (1969) Chem Phys Lett 4:155
96. Messing I, Jortner J (1977) Chem Phys 24:183
97. Born M (1920) Z Phys Chem (Muenchen, Ger) 1:221

98. Maroncelli M, MacInnis J, Fleming GR (1989) Science Washington DC, US 243:1674
99. Jortner J, Bixon M (1988) J Chem Phys 88:167
100. Biswas R, Bagchi B (1999) J Phys Chem A 103:2495
101. Rosenthal SJ, Xie X, Du M, Fleming GR (1991) J Chem Phys 95:4715
102. Kahlow MA, Kang TJ, Barbara PF (1988) J Chem Phys 90:2372
103. Walker GC, Åkesson E, Johnson AE, Levinger NE, Barbara PF (1992) Phys Chem Chem Phys 96:3728
104. Reynolds L, Gardecki JA, Frankl SJV, Horng ML, Maroncelli M (1996) J Phys Chem A 100:10337

Chapter 5
Examples of Molecular Wire Systems

In recent years, systems capable of charge and energy transport have gained tremendous scientific interest due to their applicability in electronics industry. Thus, a great variety of potential molecular wire systems have been designed, synthesized and tested. We will briefly survey a few examples to highlight the theoretical principles of charge/energy transfer through molecular systems.

Obviously, π-conjugation plays a vital role. For that reason, benzenes [1, 2], p-phenylacetylenes [3, 4] and porphyrin arrays [5–7] are well-investigated molecular structures. Besides that, more complex structures such as modified proteins and peptides [8] are at the current edge of science, especially in the field of molecular wires. Similar diversity is found when looking at the different methods of investigation. Notably, organic SAMs have been the basic building block for the majority of organic devices [9, 10]. Various types of MIM junctions [11] including single layers of molecules between aluminium and titanium/aluminium [12, 13], gold and other metal contacts [1], mechanically controllable break junctions [14], silicon adlayers [15], electrochemical break junctions [16], and SAMs sandwiched between two mercury electrodes [14, 17] have been probed in conduction measurements. At the same time, these experiments provided an extensive set of transport environments leading to a wide range of results. In this section we will, however, only concentrate on the structural aspects and overview a few examples for energy and charge transfer.

5.1 Oligo(phenylenevinylene)s

Oligo(phenylenevinylene)s (*o*PVs) are considered as a benchmark for efficient charge transport. Their composition of coplanar arranged phenyl groups that are linked by vinylene groups impose a higher rotation barrier along the phenyl ring than in, for example, *oligo*(phenyleneethynylene)s (*o*PEs) [18]. The π-conjugation, which is crucial for efficient charge transport through molecular wires, is therefore preserved by the conformational rigidity. Room temperature current–voltage (I–V) measurements in gold nanowire—SAM—gold nanowire junctions [19, 20] corroborated that conductance through the oPV molecular wire junctions is approximately one order of magnitude larger than in the corresponding oPE junctions and three orders of magnitude larger than in saturated dodecane junctions.

It seems that the same molecules tend to exhibit very different transport features which, in turn, depend on the environment. Sikes et al. [21], for example, have studied transport through oPVs and found that efficient tunneling occurs between a gold surface and an acceptor through oPV wires with a length of 28 Å. Importantly, no electrons are donated nor removed from the bridge. In contrast, Davis et al. [22, 23] investigated charge transport through a series of five DBA triads consisting of a tetracene donor, a pyromellitimide acceptor, and oPV bridges of increasing length. For short bridges, i.e. monomeric and dimeric oPVs (oPV_1 and oPV_2) they observed that a superexchange mechanism dominates the charge separation. The latter was strongly distance-dependent. However, for longer bridges, i.e. oPV_{3-5}, bridge-assisted hopping dynamics prevailed once a relatively soft distance dependence of the electron transfer resulted. The corresponding rates were found to be 3–5 times higher than in the shorter bridges.

Furthermore, Davis et al. studied the temperature dependence of charge separation and found that in none of the five cases the predictions of semi-classical ET theory do not appear to obey. In fact, they proposed that ET was "gated" by torsional motion between the tetracene donor and the first bridge phenyl ring. Support for these postulates was borrowed from the close resemblance of the activation energies for charge separation with the frequency of a known vibrational mode in 5-phenyltetracene. The overall increase of rates with increasing bridge length has been attributed to an improved planarity between donor and acceptor.

This was followed by theoretical work [24] in which density functional theory was used to shed light onto the electronic structure of oPVs and quantum dynamics calculations for investigating the ET processes. The outcome of this work supported the change in mechanism from tunneling through the bridge to its role as a real intermediate. In these studies a model was used that excluded localization of charges on the shorter bridges. The obtained rate constants agreed quite well with those observed experimentally by Davis et al. However, it was emphasized that not only torsional modes of the bridge or side groups, but also C–C stretching vibrations contribute to the higher rates in the longer bridges.

Finally, DBA systems incorporating extended tetrathiafulvalene donors (exTTF), oPV bridges, and fullerene acceptors have been constructed and investigated in Erlangen [25, 26]. Guldi et al. have found that in these exTTF–oPV–fullerene systems superexchange mediated hole transfer occurs over distances up to 50 Å with an exceptionally small attenuation factor $\beta = 0.01 \pm 0.005$ Å^{-1}. According to the authors, such excellent charge-transfer behavior is attributed to energy matching between the HOMOs of the excited donor and those of the oPV bridges and the strong electronic coupling (5.5 cm^{-1}) between the donor and acceptor moieties, mostly due to donor-bridge orbital overlap.

5.2 Oligophenylenes

Schlicke et al. [27] synthesized rod-like compounds, in which two metal-bipyridyl complexes were linked by *oligo*(phenylene)s. The longest spacer (seven phenylene units) gave rise to a metal-to-metal distance of 4.2 nm. In this work, Dexter-type energy-transfer mechanism from $[Ru(bpy)_3]^{2+}$ to $[Os(bpy)_3]^{2+}$ was established. The energy transfer mechanism is essentially temperature-dependent and decreases exponentially with an attenuation coefficient of 0.32 Å^{-1}.

In subsequent work, it was shown that the aforementioned oligomeric *para*-phenylene bridges (ph_n) act as molecular wires in the context of charge-recombination between phenothiazine and perylenediimide, when the bridge consists of at least four phenylene units [28]. The rate constants for charge recombination were obtained from the decay rate of the PDI anion. On the other hand, the corresponding DBA triads with $n = 1, 2, 3$ showed exponential distance dependence. Plotting $\log(k_{CR})$ versus r_{DA} yielded an attenuation factor of 0.67 Å^{-1}. The rate constants for $n = 4, 5$, did not lie along the linear relationship for $n = 1, 2, 3$. In fact, a trend evolves that reveals an increase of charge recombination rate with increasing bridge length. The overall dependence is depicted in Fig. 5.1. A likely rationale infers a change in charge-recombination mechanism from superexchange to thermally activated hopping upon lengthening of the bridge.

Such a change in mechanism (i.e. superexchange vs. hopping) for charge recombination strongly depends on the charge injection into the ph_n bridge. In that sense, the resulting PTZ$^{+\bullet}$–$ph_n^{-\bullet}$–PDI and/or PTZ–$ph_n^{+\bullet}$–PDI$^{-\bullet}$ states are real intermediates. Interestingly, the energies of PTZ$^{+\bullet}$–$ph_n^{-\bullet}$–PDI for $n = 1$–5 all are ≥ 3.0 eV. Thus, electron injection onto the bridge during the charge-recombination process is practically impossible. On the other hand, with increasing bridge length, the calculated energies of PTZ–$ph_n^{+\bullet}$–PDI$^{-\bullet}$ significantly decrease due to

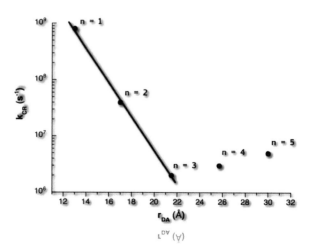

Fig. 5.1 Logarithmic plot of the charge recombination rate constant, k_{CR} versus donor–acceptor distance, r_{DA} for the PTZ–ph_n–PDI molecular wire system [29]. The best fit line through the data points for $n = 1, 2, 3$ gives $R^2 = 0.99$, $\beta = 0.67$ Å^{-1} and $k_0 = 5 \times 10^{12}$ s^{-1}

increasing conjugation length. This was corroborated by time-dependent DFT calculations. The oligophenylene band gap energy decreased from 7.3 eV for $n = 1$–3.3 eV for $n = 5$. Considering that the oxidation potential of the *oligo*-*p*-phenylene unit drops while the energy of the PTZ$^{+\bullet}$–ph_n–PDI$^{-\bullet}$ radical ion pair increases due to Coulomb destabilization, we can assume a near resonance state of PTZ–$ph_n^{+\bullet}$–PDI$^{-\bullet}$ with PTZ$^{+\bullet}$–ph_n–PDI$^{-\bullet}$ for $n = 4$ and 5. A combination of increasing electronic interactions between PTZ$^{+\bullet}$–ph_n–PDI$^{-\bullet}$ and PTZ–$ph_n^{+\bullet}$–PDI$^{-\bullet}$—as the energy gap between these two states becomes smaller—and decreasing internal reorganization energy—associated with the reduction or oxidation of the longer bridges—is responsible for this features.

5.3 Oligo(thiophene)s

Oligo(thiophene)s are a novel class of oligomeric structures. In fact, synthetic work on oligomers exceeding 10 thiophene units begun in the 1990s [29]. This was a very important step, because *oligo*(thiophene)s emerged as important molecular building blocks for developing fields such as molecular electronics. Here, liquid-crystal-display (LCD) technology should be mentioned. Since then, *oligo*(thiophene)s have been synthesized with sufficient length to bridge two nanoelectrodes made by lithographic techniques. In the year 2001, for example, Otsubo and coworkers synthesized a 72-mer [30]. The most important feature of *oligo*thiophenes is the continuous red-shift of the ground-state absorption as the chain length increases. The π-conjugation length is key to this feature [31, 32]. Nevertheless, extrapolating the absorption energy versus chain length relationship suggests that the most effective conjugation is realized in the case of 20 thiophene units—for neutral state—and around 30 units—in the oxidized state [33].

Due to the high degree of conjugation, *oligo*(thiophene)s advanced as attractive candidates for molecular bridges. For instance, Sato et al. [34] constructed hexyl-sexithiophene and methoxy-terthiophene derivatives bearing two terminal ferrocenyl groups. These served as model compounds for molecular wires. In the hexyl-sexithiophene derivative, the resultant oxidized states spread over both the ferrocene and the sexithiophene moieties. Similarly, in the methoxy-terthiophene derivative, the oxidized species spreads over the entire molecule containing the terthiophene and the other ferrocene moiety. In both cases, CT between the terminal units is inferred as it is mediated via the *oligo*(thiophene)s.

In subsequent work, Otsubo [35] investigated porphyrin–*oligo*thiophene–fullerene DBA triads. They determined the rates of intramolecular ET from the porphyrin donor to the fullerene acceptor as a function of bridge length. Quenching of the porphyrin fluorescence served as indicator for the charge separation. The resulting ET rate on nanosecond timescale decreases with increasing separation distance between donor and acceptor: 5.7×10^9 s^{-1} for $r_{DA} = 1.4$ nm, 6.2×10^8 s^{-1} for $r_{DA} = 3.0$ nm, 2.0×10^8 s^{-1} for $r_{DA} = 4.6$ nm.

Plotting $\log(k_{CS})$ verus r_{DA} yielded a linear dependence with an attenuation factor of ~ 0.11 Å^{-1}.

Nakamura et al. [36] studied photoinduced charge separation and charge recombination processes in structural analogs. They studied the kinetics in freebase porphyrin–*oligo*thiophene–fullerene (H$_2$P-nT-C$_{60}$), with $n = 4$, 8 and 12 by time-resolved fluorescence and absorption spectroscopic methods. Excitation of the H$_2$P moiety in benzonitrile and o-dichlorobenzene is succeeded by the generation of H$_2$P$^{+\bullet}$-nT-C$_{60}^{-\bullet}$. After the subsequent hole shift the outcome was the energetically stable H$_2$P-nT$^{+\bullet}$ − C$_{60}^{-\bullet}$ state. The charge separation rate for these reactions decreased with bridge length. Interestingly, the attenuation factor was found to be solvent dependent due to the stabilization of the radical ion pairs by polar solvents with very small values, i.e. 0.03 Å^{-1} in benzonitrile and 0.11 Å^{-1} in o-dichlorobenzene.

Hicks et al. [37] perturbated the electronic structure of *oligo*thiophenes by introducing strong electron-donating substituents. For that reason, 2-mesitylthio (MesS) substituents were incorporated into *oligo*thiophene bridges. The electron-donating features of the MeS substituents affected the redox properties of the *oligo*thiophenes and constrained the resulting geometry. MesS groups act as hole traps, or alternatively, as hole-hopping promoters within polymer bridges. This was corroborated by the fact, that the conjugation between the mesitylthiosulfur lone pair and the *oligo*thiophene π-system red-shifts the ground-state spectra when compared to the lack of MesS. Electrochemical measurements revealed that the MesS functionalization significantly lowers the first and second oxidation potentials of the oligomers. This lowers the coulombic barrier for charge injection into the bridge, due to the accumulation of charge density at the chain termini. Finally, these groups were shown to improve the stability of the resulting radical cations such that they evolve as more robust charge carriers.

5.4 Photonic Wires

For the sake of completeness, we should mention the function of photonic wires. A photonic wire conducts excited-state energy rather than charge from a donor to acceptor. An interesting characteristic of photonic wires is that they are operated from distance. They enable the application of light without any physical contacts to the device. Photonic wires have numerous common features with molecular wires. One of these is the extended π-conjugation. Thus, many molecular structures presented here as molecular wires can also be used as photonic wires for excited-state energy transport. For example, the first photonic wire synthesized by Lindsay and coworkers in 1994 [38] was based on conjugated porphyrin arrays. Photonic wires are capable of highly efficient multistep energy transfer across long distances [39].

References

1. Reed MA, Zhou C, Muller CJ, Burgin TP, Tour JM (1997) Science Washington DC US 278:252
2. Datta S, Tam W, Hong S, Riefenberger R, Henderson JI, Kubiak CP (1997) Phys Rev B 79:2530
3. Cygan MT, Dunbar TD, Arnold JJ, Bumm LA, Shedlock NF, Burgin TPI, Allara DL, Tour JM, Weiss PS (1998) J Am Chem Soc 120:2721
4. Seminario JM, Zacaias AG, Tour JM (1998) J Am Chem Soc 120:3970
5. Wagner RW, Lindsey JS (1994) J Am Chem Soc 116:9759
6. Hsaio J-S, Krueger BP, Wagner RW, Johnson TE, Delaney JK, Mauzerall DC, Fleming GR, Lindsey JS, Bocian DF, Donohoe RJ (1996) J Am Chem Soc 118:111816
7. Li F, Yang SY, Ciringh Y, Seth J, Martin CH, Singh DL, Kim D, Birge RR, Bocian DF, Holten D, Lindsey JS (1998) J Am Chem Soc 120:10001
8. Winkler JR, Gray HB (1992) Chem Rev Washington DC US 92:369
9. Ulman A (1996) Chem Rev Washington DC US 96:1533
10. Dubois LH, Nuzzo RG (1992) Annu Rev Phys Chem 43:437
11. Mann B, Kuhn HJ (1971) J Appl Phys 42:4398
12. Wong EW, Collier CP, Belohradsky M, Raymo FM, Stoddart JF, Heath JR (2000) J Am Chem Soc 122:5831
13. Collier CP, Wong EW, Belohradsky M, Raymo FM, Stoddart JF, Keukes PJ, Williams RS, Heath JR (1999) Science Washington DC US 285:391
14. Reed MA, Lee T (eds) (2003) Molecular nanoelectronics. American Scientific, Stevenson Ranch CA
15. Foley ET, Yoder NL, Guisinger NP, Hersam MC (2004) Rev Sci Instrum 75:5280
16. Xiao XY, Xu BQ, Tao NJ (2004) Nano Lett 4:267
17. Slowinski K, Majda M (2000) J Electroanal Chem 491:139
18. Bock CW, Trachtman M, George P (1985) Chem Phys 93:431
19. Joachim C, Gimzewski JK, Aviram A (2000) Nature 408:541
20. Reichert J, Ochs R, Beckmann D, Weber HB, Mayor M, Lohneysen HV (2002) Phys Rev Lett 88:176804
21. Sykes HD, Smalley JF, Dudek SP, Cook AR, Newton MD, Chidsey CED, Felberg SW (2001) Science Washington DC US 291:1519
22. Donhauser ZJ, Mantooth BA, Kelly KF, Bumm LA, Monnell JD, Stapleton JJ, Price DW Jr, Rawlett AM, Allara DL, Tour JM, Weiss PS (2001) Science Washington DC US 292:2303
23. Davis WB, Ratner MA, Wasielewski MR (2001). J Am Chem Soc 123:7877
24. Filatov I, Larsson S (2002) Chem Phys 284:575
25. Giacalone F, Segura JL, Martin N, Guldi DM (2004) J Am Chem Soc 126:5340
26. Martin N, Giacalone F, Segura JL, Guldi DM (2004) Synth Met 147:57
27. Schlicke B, Belser P, De Cola L, Sabbioni E, Balzani V (1999) J Am Chem Soc 121:4207
28. Weiss EA, Ahrens MJ, Sinks LE, Ratner MA, Wasielewski MR (2004) J Am Chem Soc 126:5577
29. Hoeve WT, Wynberg H, Havinga EE, Meijer EW (1991) J Am Chem Soc 113:5887
30. Otsubo T, Aso Y, Takimiya K (2001) Bull Chem Soc Jpn 74:1789
31. Sato M, Hiroi M (1994) Chem Lett 23:985
32. Bauerle P, Fischer T, Bidlingmeier B, Stabel A, Rabe J.P. (1995) Angew Chem Int Ed 34:303
33. Meier H, Stalmach U, Kolshorn H (1997) Acta Polym 48:379
34. Sato M, Fukui K, Sakamoto M, Kashiwagi S, Hiroi M (2001) Thin Solid Films 393:210
35. Otsubo T, Aso Y, Takimiya K (2002) J Mater Chem 12:2565
36. Nakamura T, Fujitsuka M, Araki Y, Ito O, Ikemoto J, Takimiya K, Aso Y, Otsubo T (2004) J Phys Chem B 108:10700

References

37. Hicks RG, Nodwell MB (2000) J Am Chem Soc 122:6747
38. Wagner RW, Lindsey JS (1994) J Am Chem Soc 116:9759
39. Heilemann M, Tinnefeld P, Sanchez Mosteiro G, Garcia Parajo M, van Hulst NF, Sauer M (2004) J Am Chem Soc 126:6514

Part III
Results and Discussion

Chapter 6
Objective

Electron and energy transfer are two of the most significant processes in chemistry and in biology. By restricting the transfer to either within a given molecule (intramolecular electron or energy transfer kinetics) or across a given molecule (single molecule transport junction conductance), it is possible to characterize both the response coefficients (rate constant or conductance) and the mechanisms for such transfers. As we have seen, the dominant mechanisms reach from coherent tunneling (i.e. large injection energy gaps, low temperatures, and short bridges) to incoherent polaron-type hopping (i.e. small injection energy gaps, higher temperatures, and longer wires). The relative importance of these two can be understood on the basis of energetic (the injection barrier vs. thermal energy) and temporal (Landauer/Buttiker contact time versus vibrational period) considerations.

In this work, we will focus on intramolecular electron and energy transfer processes. In particular, we will examine π-conjugated systems and the influence of their structural properties on electron/energy-transfer in solution. Moreover, we will concentrate on features regarding single molecules rather than properties of ensembles or higher architectures. Highly diluted solutions are used to rationalize the design of electron-transfer systems on the molecular scale. The preparation and integration of multifunctional molecular building blocks into higher architectures is believed to result in molecular tools/devices with desired functions. Molecular electronics is one of the examples. These principles are often envisaged by the expression "Molecular LEGO", referring to a molecular toolkit for the preparation of higher architectures.

Thus, the ideas behind the bottom-up approach are as simple as powerful. The general aim lies in the design of novel materials employable for molecular-scale electronics, such as molecular transistors, molecular photovoltaic applications, molecular display technology, etc. Hereby, the functions are carefully adjusted by synthetic tools to build up a certain chemical structure. In this context, we need to address the key steps/challenges in such a work-flow. With this in hands the impetus of this thesis should be illustrated.

An organic photovoltaic cell, for example, is composed of different functional components. They could be a light-harvesting unit, a molecular conductor for light-energy conversion, a redox-pair for the generation of free charge carriers (electrons or holes), and an appropriate substrate to transport the charges away from the device. Now, we would like to improve the quantum efficiency of such a cell by enhancing the amount of produced charge carriers. In this case, we would pick the component that generates charge carriers—the molecular redox pair—and probe the influence of structural modification on the overall efficiency. Consequently, we may formulate the following sequence:

1. Target synthetic modifications of the chemical structure to improve the redox properties, i.e. the energies of the resulting radical-ion pairs.
2. Experimental investigation of the modified molecules verifying the success of the structural modifications.
3. Interpretation of the experimental results using appropriate references and theory.
4. Feedback-loop, that is, reimplementation of the synthetic groups based on the results from the experimental section to further improve the investigated functionality by further chemical modifications
5. Repeating steps 1–4 until the best result is obtained.

Importantly, such a procedure is of high interdisciplinarity and requires thorough understanding of the single components making up the device and also of the mutual interactions.

The work presented in this thesis is mainly located between steps 2 and 4. It will demonstrate the control of molecular functions by means of altering the chemical structure. In particular, the focus is set on the charge-transfer properties of organic π-conjugated systems. With increasing complexity, we will add various electron donor and electron acceptor components to make novel DBA assemblies. The DBA systems will be characterized in terms of spectroscopy and theory focusing on the role of the bridge in photoinduced charge-transfer processes.

As already mentioned, features of the DBA systems cannot be understood without a significant comprehension of the individual components. In this regard, we tested the molecular wire behavior of various bridges but—for the sake of comparison—to a large extent without the variation of donors and acceptors. Only through this approach we are able to distinguish between the role of the bridge and that of the donor/acceptor pair in photoinduced charge-transfer processes.

It has already been demonstrated that wire-like behavior predominates in soluble and fully conjugated *oligo*(*para*-phenylynevinylene)s (*o*PPVs) covalently connected to a C_{60} electron acceptor and an extended tetrathiafulvalene (*ex*TTF) electron donor (C_{60}–*o*PPV–*ex*TTF) over distances of 40 Å and beyond (Fig. 6.1). Importantly, in these kind of systems an exceptionally small attenuation factor of $\beta = 0.01 \pm 0.005$ Å^{-1} is present. This is mainly due to the energy matching between the C_{60} HOMOs and the long *o*PPV-chains. Of equal importance is the strong electronic coupling between the donor (*ex*TTF) and acceptor (C_{60}). The

6 Objective

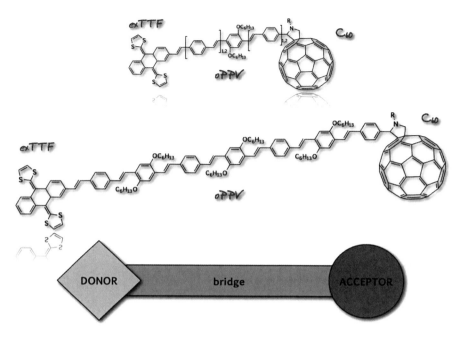

Fig. 6.1 Examples of soluble and fully conjugated *o*PPV molecular wires covalently connected to a C_{60} electron acceptor and an extended tetrathiafulvalene derivative (*ex*TTF) electron donor (C_{60}–*o*PPV–*ex*TTF). These triads exhibit wire-like behavior and electron transfer over distances of 40 Å and beyond

coupling is realized through the paraconjugation of the *o*PPVs into the *ex*TTF donor. In turn, donor–acceptor coupling constants V in $C_{60}^{-\bullet}$–*o*PPV–*ex*TTF$^{+\bullet}$ are about 5.5 cm^{-1}. This assists the rather weakly distance-dependent charge-transfer reactions [1]. As reported, *ex*TTF is a particularly interesting electron-donor molecule and has been extensively used in the preparation of electrically conducting and superconducting molecular materials [2, 3]. Despite its highly distorted structure, which shows a discernible deviation from planarity, electrically conducting charge-transfer complexes have been prepared by reacting these donors with strong electron acceptors such as fullerenes. These π-extended TTF derivatives release two electrons simultaneously (i.e. in a single quasi-reversible electron-transfer process). On the other hand, C_{60} fullerene, a three-dimensional electron acceptor, is capable of accommodating up to six negative charges. The delocalization of charges—electrons or holes—within the giant, spherical carbon framework together with its rigid, confined structure offers unique opportunities for stabilizing charged entities. Most importantly, these considerations prove that fullerenes possess small reorganization energies in electron-transfer reactions. This evolves as a definitive advantage in terms of Marcus theory, when generating long-lived charge-separated states [4].

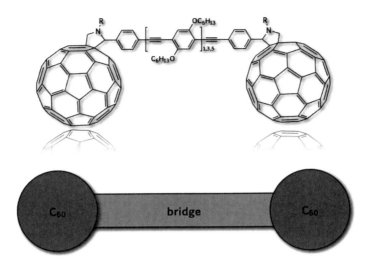

Fig. 6.2 Representative "dumbbell-type" C_{60}–oPPE–C_{60} triad with a π-conjugated oPPE bridge capable of efficient intramolecular transfer of excited state energy from oPPE to C_{60}

To conclude, we can derive several criteria that we should bear in mind when designing and studying novel molecular-wire systems, especially regarding the impact of structural parameters on the charge-transfer properties:

1. Evaluation of geometrical features and their consequences for the electronic communication between the donor and acceptor moiety.
2. Examination of the distance dependence of the charge-transfer mechanism, i.e. the impact of the length of the spacer on the charge-transfer process.
3. Analysis of the conformational flexibility of the unique components of the investigated DBA systems in terms of torsional mobility and its effect on the π-conjugation.
4. Characterization of the molecular-wire behavior, i.e. the role of the bridge in the CT reaction.

In line previous investigations of oPPV molecular wires, we have described a novel series of *oligo(para*-phenyleneethynylene) (oPPE) containing bridges with a systematically increasing length from monomer to the trimer (see Fig. 6.2). In the first place, the differences between the charge-transfer behavior of oPPV and oPPE had to be elaborated. For that reason, *ex*TTF and C_{60} was used as electron donor and electron acceptor, respectively. This work has already been carried out during my master thesis and is published elsewhere [5]. Importantly, the rational design of these novel C_{60}–oPPE–*ex*TTF systems was carried out in the background of the features that ideal molecular wires should exhibit. Hereby, π-conjugation emerged as a crucial factor for the realization of:

- small attenuation factors,
- good contacts with the electron donors and electron acceptors,
- sufficient orbital mixing of the oPPE bridge with donor and acceptor states.

Thus, the objective of this thesis is to extend the systematic characterization of photoinduced charge-transfer processes in organic conducting molecules. We considered systems, which have been explicitly designed to meet the criteria of perfect molecular conductors. By characterizing different DBA systems, we expect to apply the molecular-wire concept to understand the structural impacts on charge conducting properties in general. Furthermore, this may have important implication on solar energy conversion features, light harvesting properties, as well as quantum yields and rate constants of charge transfer.

Particularly important for photoinduced reactions is the efficient transduction of excited-state energy to the reaction centers, i.e. donor or acceptor, since this energy provides the stimulus for charge-transfer reactions. Thus, as a starting point we characterized energy-transfer properties. In general, light is considered to be absorbed by a chromophoric unit. We are dealing with highly π-conjugated bridges and, thus, most of the photons will be absorbed by the bridge generating an excited state of the bridge. If this excited-state energy is efficiently funneled to the redox centers, i.e. in our case the donor or acceptor, a subsequent charge-separation process may occur. As a result a radical ion pair is formed. In fact, such a behavior was demonstrated in the so called C_{60}–bridge–C_{60} "dumbbell" triads (Fig. 6.2), where the bridge consists of the molecular-wire of interest, namely oPPV [6], oPPE [7] and oligo(fluorene) (oFL) [8]. Here, efficient intramolecular transfer of excited-state energy from the bridge to the C_{60} was confirmed. Such dumbbell systems are considered as precursors for electron-transfer systems when replacing one of the C_{60}s by an electron donor. On the basis of characterizing photoinduced processes in such "dumbbell-type" architectures, charge-transfer processes in the corresponding DBA triads were analyzed. Particular emphasis was placed on molecular-wire behavior and conduction properties.

All systems described in this thesis contain C_{60} as electron acceptor, due to its excellent electron-accepting features. A series of different π-conjugated bridging units of variable length was investigated in terms of their molecular-wire behavior and length versus conductance dependencies. All bridges employed were covalently connected to the donors and the acceptors and exhibit full π-conjugation. As donor moieties served exTTF, ferrocene (Fc) and porphyrins, i.e. free base (H_2P) and metallated (e.g. ZnP). The particular specific structures of the investigated compounds will be presented in the next chapters.

In general, a thorough spectroscopic study, as routinely carried out in the group of Prof. Dr. Dirk M. Guldi by means of steady-state emission/absorption measurements and time-resolved techniques in numerous solvents, sheds light onto the photophysical processes following photoexcitation of these systems. Equally, a detailed description of the employed spectroscopic methods will be given in the next sections.

Finally, quantum mechanical calculations in collaboration with Prof. Dr. Timothy Clark were employed to corroborate the experimental findings and provide additional insight into the structural and electronic properties of the compounds. Furthermore, molecular modeling was used to simulate the charge-transfer processes in vacuo and various solvents.

In the next chapters the subsequent issues will be elaborated:

1. A description of the experimental details applied throughout the photophysical characterization, i.e. spectroscopic methodology and apparatus setups (Chap. 7).
2. Characterization of photoinduced energy-transfer processes and mechanisms (Chap. 8).
3. Characterization of photoinduced electron-transfer processes and mechanisms of DBA systems involving (Chap. 9)

    1. *oligo*(phenyleneethynylene) (*o*PE) molecular wires (Sect. 9.1),
    2. *oligo*(fluorene) (*o*FL) molecular wires (Sect. 9.2),

4. Comparative conclusions (Chap. 10).

References

1. Giacalone F, Segura JL, Martin N, Guldi DM (2004) J Am Chem Soc 126:5340
2. Becher J, Zhan-Ting L, Blanchard P, Svenstrup N, Lau J, Brondsted Nielsen M, Leriche P (1997) Pure Appl Chem 69:465
3. Yamada J, Sugimoto T (eds) (2004) TTF Chemistry. Springer Berlin Heidelberg Kodansha Scientific Ltd., Tokyo Japan
4. Imahori H, Sakata Y (1999) Eur J Org Chem 10:2445
5. Atienza C, Martín N, Wielopolski M, Haworth N, Clark T, Guldi DM (2006) Chem Commun (Cambridge UK) 30:3202
6. Giacalone F, Segura JL, Martín N, Ramey J, Guldi DM (2005) Chem–Eur J 11:4819
7. Atienza C, Insuasty B, Seoane C, Martín N, Ramey J, Guldi DM (2005) J Mater Chem 15:124
8. Pol Cvd, Bryce MR, Wielopolski M, Atienza-Castellanos C, Guldi DM, Filippone S, Martín N (2007) J Org Chem 72:6662

Chapter 7
Instruments and Methods

7.1 Photophysics

All systems were probed in steady-state fluorescence, absorption and time resolved emission lifetime studies at room temperature. Additionally, time resolved femtosecond transient absorption and nanosecond laser flash photolysis measurements were carried out.

The following experimental setups and instruments were used for the various photophysical techniques.

7.1.1 Absorption Spectroscopy

All UV/vis spectra were recorded on a *Varian* Cary 50 Scan spectrophotometer and a *PERKIN ELMER* UV/vis Spectrometer Lambda 2 (double beam) in solution. Absorption maxima λ_{max} are given in nm. 0.4 cm quartz cuvettes were used for all measurements.

7.1.2 Steady-state Emission

Steady state fluorescence studies were carried out on a Fluoromax 3 (*Horiba*) instrument in solution and all spectra were corrected for the instrument response. Emission maxima λ_{max} are given in nm. 0.4 cm quartz cuvettes were used for all measurements.

7.1.3 Time-resolved Emission

Fluorescence lifetimes were measured with a Laser Strobe Fluorescence Lifetime Spectrometer (*Photon Technology International*) with 337 nm laser pulses

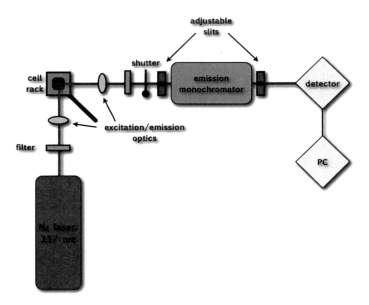

Fig. 7.1 Schematic representation of the Laser Strobe system for fluorescence lifetime measurements. A nitrogen laser is used as excitation source. The time resolution is limited to approximately 0.1 ns

(800 ps) from a nitrogen laser fiber-coupled to a lens-based T-formal sample compartment equipped with a stroboscopic detector. Details of the Laser Strobe systems are described on the manufacture's web site. Fluorescence lifetimes were measured at the emission maximum. A schematic representation of the setup is given in Fig. 7.1. The results were elaborated with the provided *Felix* software. 0.4 cm quartz cuvettes were used for all measurements.

7.1.4 Femtosecond Transient Absorption Spectroscopy

The femtosecond transient absorption studies were performed with 387 nm laser pulses (1 khz, 150 fs pulse width) from an amplified Ti:Sapphire laser system (Model CPA 2101, *Clark-MXR Inc.*). A NOPA optical parametric converter was used to generate ultrashort tunable visible pulses from the pump pulses. The apparatus is referred to as a two-beam setup, where the pump pulse is used as excitation source for transient species and the delay of the probe pulse is exactly controlled by an optical delay rail. As probe (white light continuum), a small fraction of pulses stemming from the CPA laser system was focused by a 50 mm lens into a 2-mm thick sapphire disc. A schematic representation of the setup is given below in Fig. 7.2. 2.0 mm quartz cuvettes were used for all measurements.

7.1 Photophysics

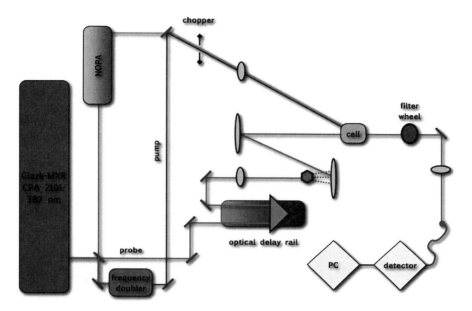

Fig. 7.2 Two-beam experimental setup for femtosecond transient absorption studies using a white light continuum. A commercially available CPA 2101 laser system delivers the pulses. Ultrashort tunable visible pulses are obtained by the NOPA optical parametric converter. A chopper wheel is used to cut every second pump pulse in order to compare the signal with and without the pump. The white light continuum is generated by a sapphire disc. The time delay between the pump and probe pulses is adjusted by the optical delay rail

7.1.5 Nanosecond Laser Flash Photolysis

Nanosecond laser Flash Photolysis experiments were performed with 355 and 532 nm laser pulses from a *Brilland-Quantel* Nd:YAG system (5 ns pulse width) in a front face (VIS) and side face (NIR) geometry using a pulsed 450 W XBO lamp as white light source. Similarly to the femtosecond transient absorption setup, a two beam arrangement was used. However, the pump and probe pulses were generated separately, namely the pump pulse stemming from the Nd:YAG laser and the probe from the XBO lamp. A schematic representation of the setup is given below in Fig. 7.3. 0.5 cm quartz cuvettes were used for all measurements.

7.2 Chemicals

All chemicals and solvents were purchased from *Sigma Aldrich* in spectrophotometric grade and used without further purification.

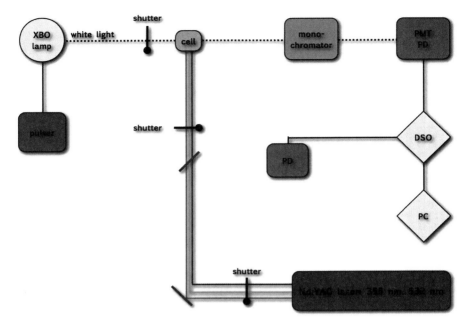

Fig. 7.3 Experimental setup for the nanosecond laser Flash Photolysis with a white light continuum. A *Brilland-Quantel* Nd:YAG laser delivers the fundamental pulses (355 and 532 nm). A pulsed XBO lamp is used as white light source. The laser signal is split in order to trigger the digital storage oscilloscope (DSO) utilizing a second photodiode (PD). Two separate detection units in different geometries—photomultiplier (PMT) in front face and a PD in side face—detect the signal in the UV/vis and NIR region, respectively. The monochromator is operated by a standard PC

The investigated compounds were synthesized and characterized by various synthetic groups and are described elsewhere (see publication list). They were used without further processing or purification in solutions of various solvents as mentioned in the text.

7.3 Molecular Modeling

Molecular properties in the electronic ground state were computed using density functional theory (DFT) and the restricted Hartree–Fock formalism with various semi-empirical Hamiltonians (see text for details). Semi-empirical excited-state calculations used a full configuration interaction (C.I.) expansion with an active window of a defined number of highest occupied and lowest unoccupied molecular orbitals. Test calculations with a singles only C.I. including up to 30 active orbitals were performed prior to any these calculations. Solvent effects were simulated using single-point calculations on the gas-phase optimized geometries with the

7.3 Molecular Modeling

self-consistent reaction field solvation (SCRF) models considering solvent-excluded surfaces with atomic radii equal to 120% of those given by Bondi [1].

Theoretical calculations within the DFT approach were carried out by using the D.02 revision of the *Gaussian 03* suite of programs [2]. Semi-empirical calculations were performed using the *VAMP* 10.0 package [3]. Local molecular properties, e.g. electron affinity maps and electrostatic potentials were computed with *Parasurf 07* [4–6] and visualized using *Tramp 1.1d* [7]. In some cases, additional force field calculations were accomplished with the Forcite package as implemented in *Materials Studio 4.2* [8]. All molecular structure and orbital representations were generated with the *Chimera* visualization program [9] or *Materials Studio 4.2* [8].

References

1. Bondi AJ (1964) J Phys Chem A 68:441
2. Frisch MJ, Trucks GW, Schlegel HB, Scuseria GE, Robb MA, Cheeseman JR, Montgomery JA Jr, Vreven T, Kudin KN, Burant JC, Millam JM, Iyengar SS, Tomasi J, Barone V, Mennucci B, Cossi M, Scalmani G, Rega N, Petersson GA, Nakatsuji H, Hada M, Ehara M, Toyota K, Fukuda R, Hasegawa J, Ishida M, Nakajima T, Honda Y, Kitao O, Nakai H, Klene M, Li X, Knox JE, Hratchian HP, Cross JB, Bakken V, Adamo C, Jaramillo J, Gomperts R, Stratmann RE, Yazyev O, Austin AJ, Cammi R, Pomelli C, Ochterski JW, Ayala PY, Morokuma K, Voth GA, Salvador P, Dannenberg JJ, Zakrzewski VG, Dapprich S, Daniels AD, Strain MC, Farkas O, Malick DK, Rabuck AD, Raghavachari K, Foresman JB, Ortiz JV, Cui Q, Baboul AG, Clifford S, Cioslowski J, Stefanov BB, Liu G, Liashenko A, Piskorz P, Komaromi I, Martin RL, Fox DJ, Keith T, Al-Laham MA, Peng CY, Nanayakkara A, Challacombe M, Gill PMW, Johnson B, Chen W, Wong MW, Gonzalez C, Pople JA (2004) Gaussian 03 Revision D02. Gaussian Inc, Wallingford CT
3. Clark T, Alex A, Beck B, Burkhardt F, Chandrasekhar J, Gedeck P, Horn A, Hutter M, Martin B, Rauhut G, Sauer W, Schindler T, Steinke T (2003) VAMP 10.0. Erlangen
4. Clark T, Lin H Jr, Horn AHC (2007) Parasurf 07. Cepos InSilico Ltd., Ryde, UK. http://www.ceposinsilico.com
5. Ehresmann B, Martin B, Horn AHC, Clark T (2003) J Mol Mod 9:342
6. Clark T (2004) J Mol Graphics Modell 22:519
7. Lanig H, Koenig R, Clark T (2005) Tramp 1.1d Erlangen
8. Accelrys Software Inc. (2007) Forcite Modeling Environment Release 42. Accelrys Software Inc, San Diego
9. Pettersen EF, Goddard TD, Huang CC, Couch GS, Greenblatt DM, Meng EC, Ferrin TE (2004) UCSF Chimera - a visualization system for exploratory research and analysis, vol 25. J Comput Chem

Chapter 8
Energy Transfer Systems

For a thorough characterization of processes resulting from the photoexcitation of organic π-conjugated molecules, we should focus on systems capable of energy transfer reactions. In such, the absorption of light and the subsequent deactivation of the excited state transfers excited state energy to an acceptor moiety rather than resulting in the generation of radical ion pairs. Understanding these processes helps in the identification of the light-harvesting properties of the π-conjugated oligomers, referred to as molecular wires.

Energy transfer reactions in systems containing the herein investigated molecular wire structures have already been investigated in Erlangen. Two well-characterized examples, which were investigated within the scope of this thesis will be presented in more detail. This lines out the characteristic features of photoinduced energy transfer reactions.

8.1 Linking two C_{60} Electron Acceptors to a Molecular Wire

In Chap. 6 we have already discussed the concept of the so called C_{60}–wire–C_{60} "dumbbell" conjugates as simple reference systems for wire-like π-conjugated systems. Understanding the deactivation mechanisms (i.e. rates and quantum yields) in such systems is essential to identify potential molecular wire candidates for their integration into donor-bridge–acceptor conjugates. In other words, probing and quantifying energy-transfer characteristics in such systems is essential for understanding the processes that are likely to occur when one of the C_{60} acceptors is replaced by a molecular donor.

8.1.1 C_{60}–oPPE–C_{60}—A Representative Example for Efficient Energy Transfer

In view of the molecular-wire properties of *oligo*-phenyleneethynylenes (*o*PEs), C_{60}–wire–C_{60} conjugates **3** have been investigated, where two C_{60}s were attached

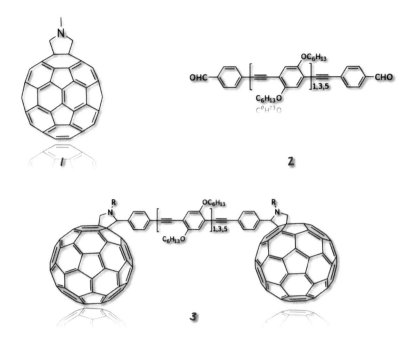

Fig. 8.1 C_{60}–*oligo*(para-phenyleneethynylene)$_n$–C_{60} "dumbbell" triads and their building blocks as reference compounds

to rigid *oligo*(para-phenyleneethynylene) (*o*PPE) moieties with oligomeric units that ranged from 1 to 7 (Fig. 8.1) [1].

Atienza et al. studied the excited state properties of these triads by means of fluorescence and transient absorption spectroscopic measurements. Interestingly, in this type of systems the bridge reveals a weak donating character when compared to the strong accepting features of fullerenes. This was highlighted in electrochemical studies, where an amphoteric redox behavior was registered. Another conclusion of the electrochemical measurements is that no electronic communication is present between the bridge and the C_{60} moieties in the ground state. This is in good agreement with the assumption that the bridging unit is just an integrative building block linking the two C_{60} termini. The length of the bridge fails to impede the reduction properties of the fullerene. A likely rationale is the presence of two sp^3 carbon atoms that are placed between C_{60} and the π-conjugated oligomer. Thus, one may conclude that the bridge is electronically separated from the C_{60}s.

The lack of significant electronic interactions between the electroactive constituents in the ground state prompts to the excited state characteristics. This has been accomplished by various photophysical methods. A short summary of the results will be presented in this chapter to set the tone for the following energy transfer discussions.

8.1 Linking two C_{60} Electron Acceptors to a Molecular Wire

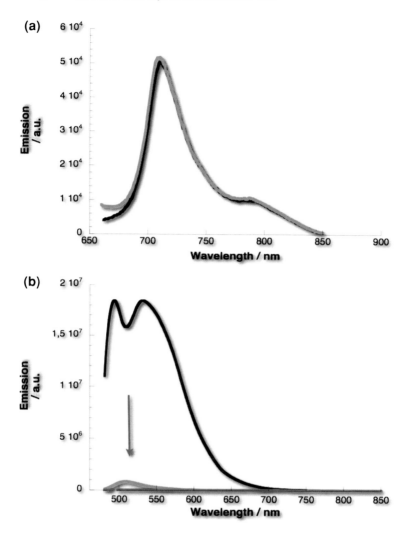

Fig. 8.2 a Fullerene emission of the C_{60}-reference (*N*-methylfulleropyrolidine) **1** (*black line*) and the C_{60}–*o*PPE–C_{60} triads **3** (*orange line*)—in toluene with matching absorption at the 390 nm excitation wavelength (i.e. $OD_{390nm} = 0.2$). **b** Fluorescence spectra of **2** (*black line*), **3** (*orange line*), and **1** (*red line*) displaying the quenching of the *o*PPE fluorescence in the triads in respect to the *o*PPE reference compound—in toluene with matching absorption at the 390 nm excitation wavelength (i.e., OD = 0.2)

Within the context of the current systems, only the weak fluorescing features of C_{60} that are maximized at 710 nm should be mentioned explicitly. A representative emission spectrum is illustrated in Fig. 8.2a. Regarding, the phenylene–acetylene building blocks, the major features include strong visible light fluorescence with quantum yields close to unity. Similarly to the absorption maxima, the fluorescence maxima depend on the length of the π-conjugated *oligo*-phenyleneethynylenes and reach from 400 to 490 nm (Fig. 8.2b).

This length dependence is also associated with fluorescence related features. Lifetimes and quantum yields increase with increasing chain length. Due to the strong absorptions of the *o*PPE building blocks in the visible, i.e. 300–600 nm, and the marginal C_{60} absorption cross section, which is only competitive in the UV region of the spectrum, visible light excitation reaches exclusively the *o*PPE bridge units. A simple comparison between the strongly emissive features of the *o*PPE building block references **2** (i.e. lacking C_{60}) in the 500–700 nm range and the emissive features of the triads provided insight into the intramolecular deactivation processes. It turned out, that in the C_{60}–*o*PPE–C_{60} conjugates the strong *o*PPE emission is drastically quenched. This has been attested to an almost instantaneous and quantitative deactivation of the *o*PPE building blocks (Fig. 8.2b). On the other hand, fluorescence of C_{60} in the range between 700 and 800 nm is a qualitative match of that of the C_{60} reference fluorescence—see Fig. 8.2a. In conclusion, the quenching of *o*PPE and the presence of C_{60} fluorescence implied an exothermic transduction of singlet excited state to the low-lying C_{60} excited state. Furthermore, excitation spectra, in which the C_{60} emission at 710 nm was probed as a function of excitation wavelength, resembled the ground-state absorption. Analyzing the *o*PPE features implies that the origin of the excited state energy is unquestionably that of the conjugated π-conjugated system. The only rationale for such steady-state emission behavior is an efficient intramolecular excited state energy transfer from the photoexcited *o*PPE bridges to C_{60}. The influence of solvent polarity on the emission features is negligible. This serves further an independent, indirect verification of excited state deactivation, which is governed by energy transfer. After the completion of the singlet–singlet energy transfer, the fullerene singlet

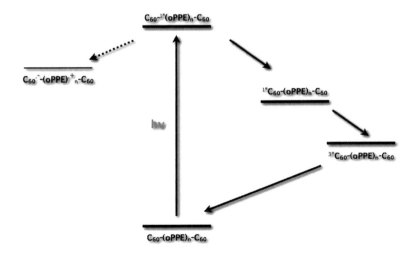

Fig. 8.3 Schematic illustration of the reaction pathways in photoexcited **3**

excited state can only decay to the triplet excited state and finally to the singlet ground state. This is schematically sketched in Fig. 8.3. In order to identify the photoproducts, transient absorption measurements upon nanosecond laser excitation at 308, 337 and 355 nm were employed. The spectroscopic and kinetic analyses suggested the formation of only one photoproduct, namely the C_{60} triplet excited state. Spectroscopic proof for the latter are transient maxima at 360 and 700 nm.

In summary, the deactivation of the photoinduced excited states in the C_{60}–oPPE–C_{60} triads is dominated by efficient intramolecular energy transfer, which thermodynamically (i.e. larger energy gap) and mechanistically (i.e. efficient dipole–dipole interactions) prevails. These rapid energy-transfer processes occur although no significant interactions between oPPE and C_{60} are present in the ground state, i.e. bridge and acceptor are two independent entities.

8.1.2 Energy Transfer in C_{60}–oligo(fluorene)–C_{60}

With increasing interest in *oligo-* and *poly*fluorenes as suitable materials for the preparation of molecular electronic devices, such as organic light emitting diodes (OLEDs), we have focused on their optical and electronic properties in terms of molecular wire behavior. In line with the above mentioned results obtained on C_{60}–oPPE–C_{60} systems, similar architectures were synthesized utilizing *oligo*(fluorene) (oFL) bridges [2]. Important is in that context, that Wasielewski et al. [3] have shown that the length of oFLs used as a bridge component between a donor and acceptor moiety does not affect the energies of the relevant bridge states. In other words, the oxidation potential of *oligo*(fluorene) bridges is invariant upon changing the length of the bridge. These features are attributed to the localization of charges on the two terminal fluorene units in $(oFL)_n$. As a consequence, outstanding electronic coupling between the bridge and the redox-active moieties must be present. Motivated by these findings, we have chosen to investigate the electronic properties of oFLs as implemented into C_{60}–oFL–C_{60} conjugates and further into DONOR–oFL–C_{60} architectures (see Sect. 9.2).

For that reason, novel C_{60}–oFL **5** and C_{60}– oFL–C_{60} **6** (Fig. 8.4) bearing *oligo*(fluorene) moieties of different lengths, in particular dimer, trimer and pentamer, have been synthesized and investigated in terms of their photophysical behavior.

8.1.2.1 Molecular Modeling

In order to assert the charge localization on the termini of the oFLs, semi-empirical calculations have been employed on the positively charged *oligo*(fluorene) dimer, trimer and pentamer. The geometry-optimized structures (AM1 [4]) have been investigated in their neutral and positively-charged states. Figure 8.5 represents the

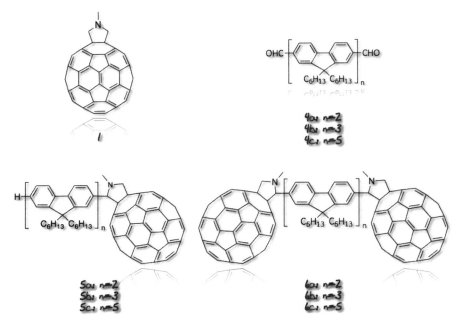

Fig. 8.4 C_{60}–*oligo*(fluorene)$_n$ dyads and C_{60}–*oligo*(fluorene)$_n$–C_{60} "dumbbell" triads with their building blocks as reference compounds

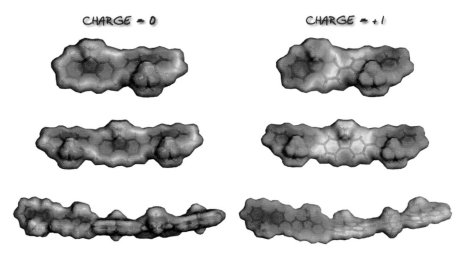

Fig. 8.5 Representation of the electrostatic potential of **4a–c** and its redistribution upon charging the *o*FL moieties. In all cases the negative charge (*blue*) is localized at the termini of the *o*FLs relative to the neutral species

8.1 Linking two C$_{60}$ Electron Acceptors to a Molecular Wire

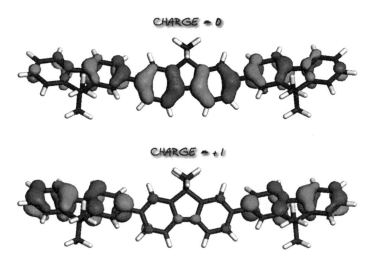

Fig. 8.6 HOMO orbital redistribution in **4b** upon charging the molecule. The electron density is shifted to the termini of the charged species

electrostatic potential of the neutral and the charged species. Independent on the length of the oligomer the charges localize at the termini of the oFL chains. Important for our consideration is that the calculated ionization potentials (AM1) do not vary with the bridge length (177 kcal/mol for the dimer, 174 kcal/mol for the trimer, 169 kcal/mol for the pentamer). Furthermore, the analysis of the frontier orbitals, as depicted in Fig. 8.6 for the trimer, corroborates the electron donor properties of the oFLs, as the HOMOs in the charged species are shifted and spatially separated relative to the neutral state. The gain of symmetry in the arrangement of the HOMOs accounts for facile stabilization of the charged-species.

When connecting oFLs to two C$_{60}$, the weak but noticeable donor character of the linker unveils. Semi-empirical configuration interaction (CI) calculations revealed that, similarly to oFL lacking C$_{60}$, the positive charge is localized at the terminal sites of the bridge rather than somewhere else (Fig. 8.7). According to these findings, we can assure noticeable electronic coupling between oFL and C$_{60}$ and the preservation of a constant oFL oxidation potential upon increasing their length (HOMO and LUMO are spatially separated—Fig. 8.8).

The spatial separation of the positive and negative charge at oFL and C$_{60}$, respectively, and the lack of electronic communication between them in the ground state imply that charge-separation features are rather negligible. This is particularly evident when comparing them with similar DONOR-oFL–C$_{60}$ compounds. In fact, the charge-separated states found in these calculations exhibit only a low change of dipole moment of approximately 45 Debye. Please compare that to corresponding DONOR–oFL–C$_{60}$ architectures (>100 Debye). Ionization potentials of the bridges are lower in the presence of C$_{60}$ than in its absence.

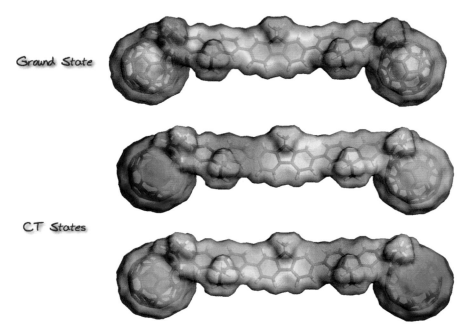

Fig. 8.7 Representation of the electrostatic potential of **6b** in its ground state and its two charge-transfer (CT) states. Again, the positive charge (*red*) is localized at the termini of the *o*FL bridge relative to the neutral species

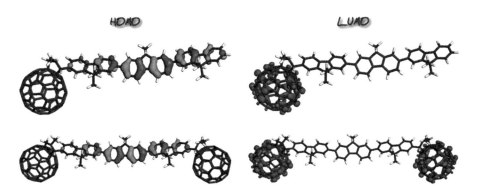

Fig. 8.8 HOMO/LUMO orbital schemes of **5b** and **6b** representing the spatial separation between donor and acceptor moiety

The excellent electron-accepting features of C_{60} are responsible for this change, which is independent on the bridge length, namely 95 kcal/mol for **5a,b**, 96 kcal/mol for **5c**, 97 kcal/mol for **6a,b**, and 96 kcal/mol for **6c**.

8.1.2.2 Photophysics

Like the C_{60}–oPPE–C_{60} conjugates, C_{60}–oFL and C_{60}–oFL–C_{60} conjugates give rise in electrochemical measurements to an amphoteric redox behavior, that is, the presence of three quasireversible reduction waves, which correspond to the three reduction steps of C_{60}. These features were found in all oligomers and imply a lack of electronic communications between C_{60} and oFL in the ground state of **5a–c** and **6a–c**. Additionally, a donor character of oFLs is seen. This is subject to cathodic shifts when going from the monomer to the dimer, trimer and pentamer, which is attributed to the extension of π-conjugation upon increasing oligomer length.

These findings were corroborated by the absorption characteristics of the conjugates, which failed to be a superposition of the individual building blocks.

Inspecting the steady-state fluorescence of the reference compounds, **4a–c**, (Fig. 8.9), displays the strong emission features throughout the visible region, i.e. 400–600 nm. Similar to C_{60}–oPPE–C_{60}, in **5a–c** and **6a–c** this oligomer emission is markably quenched, namely by a factor of 10^3 in toluene, for instance. In general, the quenching is higher in the triads than in the diads. Interestingly, the fluorescence pattern of the oFLs is preserved, despite the presence of one or two fullerenes. This is indicative for the lack of electronic communication found by electrochemical and absorption measurements. In the near infrared region (650–800 nm) the C_{60} (*N*-methylfulleropyrolidine, **1**) fluorescence pattern is seen for **5** and **6**. At this point, it should be stressed that similarly to the C_{60}–oPPE–C_{60} conjugates, an excitation at 370 nm directs the light almost quantitatively to the *oligo*(fluorenes) and not to C_{60}.

As a matter of fact, we may assume that the singlet excited state energies of all oFLs (2.70 eV) are quantitatively transduced to C_{60} (1.76 eV). This is followed by an efficient intersystem crossing to yield the fullerene triplet excited state. The energy transfer reaction was quantified by comparing the C_{60} fluorescence of **5** and **6** in, for example, toluene with that of C_{60}-reference **1** (6.0×10^{-4}). Herby, the latter served as an internal reference when exactly the same experimental conditions are applied. Quantum yields close to 6.0×10^{-4} speak for a quantitative energy transfer in all the tested systems.

Independent confirmation for the proposed energy-transfer mechanism stems from femto- (387 nm) and nanosecond (355 nm) transient absorption measurements, where both the fullerene and oFL units exhibit similar absorptions. Considering the "naked" oligomer reference compounds **4a–c**, photoexcitation generates instantaneously meta-stable singlet excited state transients characterized by a ground-state bleaching in the 400–450 nm region and a new transient absorption in the 600–1200 nm region (Fig. 8.10). These processes occur on a time-scale of less than a ps (Fig. 8.11). Furthermore, it was found, that the transient maxima depend on the length of the oligofluorene unit: 670 nm (dimer, **4a**), 725 nm (trimer, **4b**) and 752 nm (pentamer, **4c**). In all oligomers, the product of the decay (i.e. 0.8 ns) is the corresponding triplet excited-state of the oFLs.

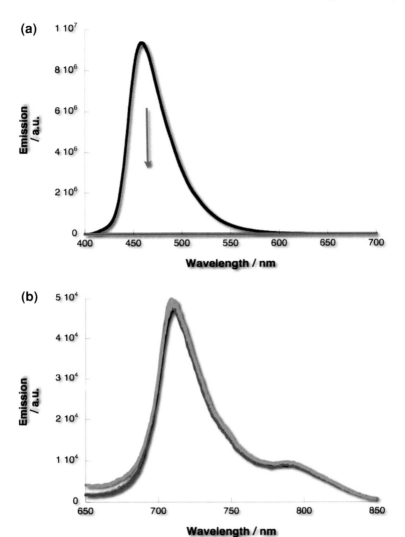

Fig. 8.9 **a** Fluorescence spectra of **4a** (*black spectrum*), **6a** (*red spectrum*), **6b** (*orange spectrum*) and **6c** (*pink spectrum*), in THF, with matching absorption of 0.2 at the 345 nm excitation wavelength, **b** fluorescence spectra of **1** (*black spectrum*), **6a** (*red spectrum*), **6b** (*orange spectrum*) and **6c** (*pink spectrum*) in THF, with matching absorption of 0.2 at the 370 nm excitation wavelength

Upon excitation of both the C_{60}–*o*FL dyads **5a–c** and C_{60}–*o*FL–C_{60} triads **6a–c** transient features develop that agree with these seen in the *o*FL references, **4a–c**, namely transient bleaching below 450 nm and transient maxima around 700 nm (Fig. 8.12). Thus, such observations attest—in close agreement with the ground

8.1 Linking two C_{60} Electron Acceptors to a Molecular Wire

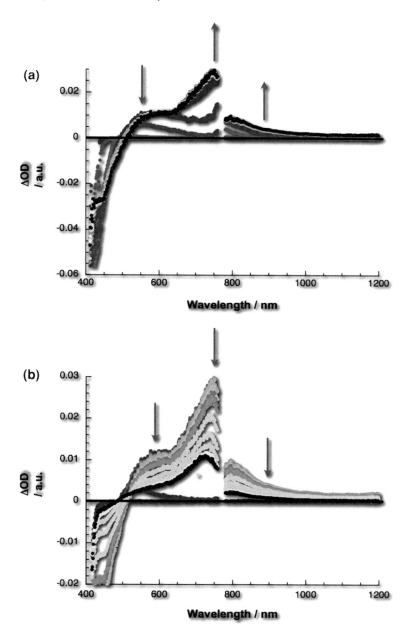

Fig. 8.10 a Differential absorption spectra (visible and near-infrared) obtained upon femtosecond flash photolysis (387 nm) of **4b** in nitrogen-saturated THF solutions with several time delays between 0 and 20 ps at room temperature and **b** between 0 and 1600 ps at room temperature

Fig. 8.11 Time–absorption profiles of the spectra shown above at 655 and 745 nm, monitoring the formation of the singlet excited state

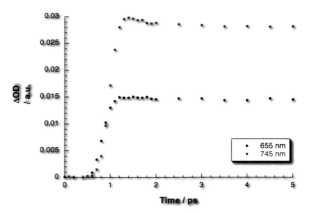

state absorption at the excitation wavelength of 387 nm—the selective excitation of the oFL moieties.

In contrast to the reference compounds **4a–c**, the singlet excited state deactivates in the dyads and triads much faster than in the oFL references with rate constants typically on the order of $\sim 10^{12}$ s^{-1} (Fig. 8.13). These values match the quantitative quenching of the oFL fluorescence in **5** and **6**. A comparison between the decay rates of the conjugates carrying one C_{60} or two C_{60} moieties, reveals a 2-fold acceleration of the *oligo*(fluorene) deactivation in the latter. Conclusively, placing two C_{60}s instead of one onto the oFL backbone significantly accelerates the excited state decay.

Toward the end of the decay of the oligofluorene singlet absorption, only the transient maximum of the C_{60} singlet excited state (880 nm) is seen. This verifies the successful transduction of singlet excited state energy from the oFL moieties to C_{60}. On a time scale of 3 ns the fullerene singlet excited state intersystem crosses to the corresponding triplet excited state. Spectroscopically, the development of a new transient maximum at 700 nm corresponds to the C_{60} triplet excited state (Fig. 8.14a). Importantly, the kinetics of the singlet decay and the growth of the triplet match each other reasonably well, as evidenced by Fig. 8.14b.

Complementary nanosecond experiments ratify the photosensitization of the triplet features in **5** and **6** (Fig. 8.15). In the absence of molecular oxygen multi-exponential decay kinetics point to a fairly complex deactivation scheme.

Incontrovertible, an efficient transduction of singlet excited-state energy transfer prevails from the photoexcited oFL to the energy-accepting fullerenes in the C_{60}–oFL dyads and C_{60}–oFL–C_{60} triads. Such a reactivity is perfectly in line with the previously described behavior observed in corresponding C_{60}–oPPE–C_{60} systems. Importantly, no spectral evidence has been found that would suggest a competing electron-transfer reaction. The latter may evolve, for instance, from the energetically high lying singlet excited states of the oligofluorenes. A schematic representation of the possible reaction pathways is given in the energy diagram of Fig. 8.16.

8.1 Linking two C_{60} Electron Acceptors to a Molecular Wire

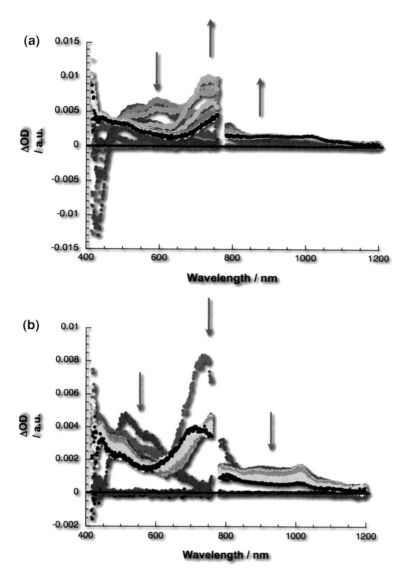

Fig. 8.12 **a** Differential absorption spectra (visible and near-infrared) obtained upon femtosecond flash photolysis (387 nm) of **6b** in nitrogen-saturated THF solutions with several time delays between 0 and 2 ps at room temperature and **b** between 0 and 1600 ps

Thus, addressing DONOR–$(o\text{FL})_n$–C_{60} ensembles utilizing strong electron donor groups advanced as a promising task to probe the role of *oligo*(fluorene) molecular wires into donor–acceptor couples. The corresponding results will be presented further in this thesis.

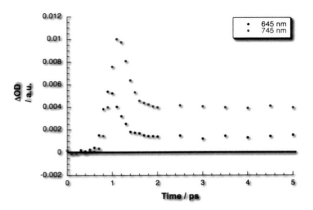

Fig. 8.13 Time–absorption profiles of the spectra shown above at 645 and 745 nm, monitoring the formation of the singlet excited state

8.2 Tunable Excited State Deactivation

An interesting example of through-space energy-transfer in systems involving C_{60} will be presented as another example of photoinduced excited state deactivation. Particularly important is the tunability by designing the photoactive units in such energy transfer systems. For that reason, together with the synthetic chemist's group in Erlangen, we have developed an approach to establish the energy-accepting features of fullerenes in light-harvesting antenna systems composed of dendritic structures. Such dendrimer approach provides a very high degree of tunability of the light-harvesting properties and, importantly, of the energy levels of the particular building blocks. Herein, we will show how a non-covalent binding motif, namely hydrogen bonding, is utilized for assembling hybrids that undergo photoinduced energy transfer reactions. Among the many supramolecular binding motifs that are available, hydrogen bonding is considered as the structurally most potent one. In this context, hydrogen bonding offers great control over fine-tuning of the complexation strength, regulation of the electronic coupling and its impact on electron- and energy-transfer processes.

Inspired by a modular concept toward supramolecular heme-protein models, chromophoric and non-chromophoric dendrons, which are interchangeable at the tin-tetraphenyl-porphyrin (SnP) platform containing two axial hydrogen bonding bridging sites, have been synthesized. The binding sites are referred to as "Hamilton receptors" due to their ability to bind barbiturates and their derivatives through six hydrogen bonds. Such a binding scheme was introduced in 1988 by Hamilton et al. [5]. Energy transfer between porphyrins and fullerenes and the investigation of successive fluorescence quenching in 1:2 complexes between the SnP and a fullerene containing a dendron in its second generation will be presented (Fig. 8.17) [6]. The binding between the two building blocks in the supramolecular architecture is realized via strong six-point hydrogen bonding with the complementary "Hamilton receptors" of SnP. The self-assembly of the SnP **8** with

8.2 Tunable Excited State Deactivation

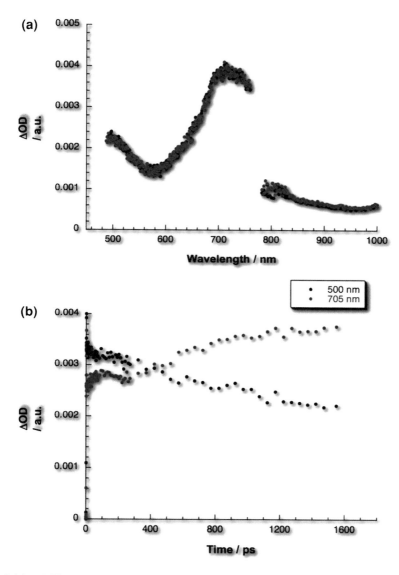

Fig. 8.14 **a** Differential absorption spectra (visible and near-infrared) obtained upon femtosecond flash photolysis (387 nm) of **6b** in nitrogen-saturated THF solutions with a time delay of 3000 ps at room temperature, indicating the fullerene Triplet–triplet features. **b** Time–absorption profiles of the spectra shown above at 500 and 705 nm, monitoring the decay of the singlet excited-state and formation of the triplet excited state

the depsipeptide fullerene ligand **7** was investigated by NMR spectroscopic methods and a 1:2 complex formation was established instigated by hydrogen bonding [7].

Fig. 8.15 Differential absorption spectra (visible and near-infrared) obtained upon nanosecond flash photolysis (355 nm) of **6b** (2.0 × 10^{-5} M) in nitrogen-saturated THF solutions with a time delay of 100 ns at room temperature, indicating the fullerene triplet–triplet features

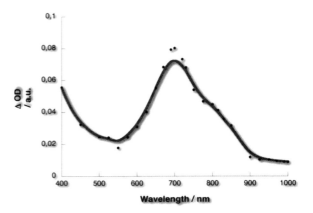

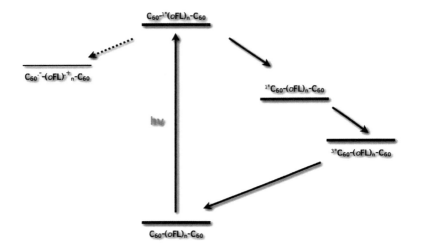

Fig. 8.16 Schematic illustration of the reaction pathways in photoexcited **6**

8.2.1 Photophysics

Titration assays, in which complex formation of **8:7** was probed in the presence of variable concentrations of **7** provided insight into the excited state interactions, namely the deactivation of the singlet excited state of the porphyrin via energy transfer to the fullerene core. In particular, adding increasing amounts of **7** to the SnP solution results in substantial quenching of the porphyrin emission centered at 610 and 665 nm. As illustrated in Fig. 8.18, the quenching is exponential and depends exclusively on the added concentration of **7**, which points to the

8.2 Tunable Excited State Deactivation

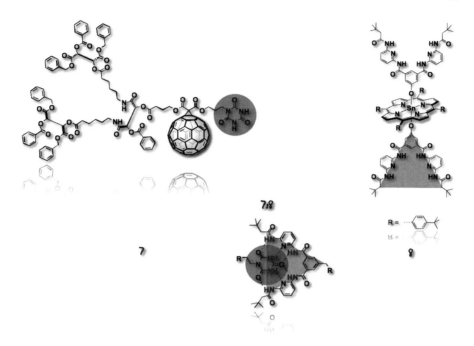

Fig. 8.17 *Upper part*: structures of **7** and the tin-tetraphenyl-porphyrin (SnP) **8**. *Lower part*: schematic representation of the hydrogen bonding motif

successful formation of **8:7₂**. The nonlinear relationship gave rise to association constants for the first and second binding steps, that are with $\log K_1 = 3.6$ M^{-1} and $\log K_2 = 6.9$ M^{-1} quite strong. Notably, at the end point of the titration the porphyrin fluorescence is nearly quantitatively quenched, which is attributed to the quantitative formation of **8:7₂**. Instead, a fluorescence pattern (i.e., *0–0 emission around 700 nm) evolved that is reminiscent of that of **7**. Interestingly, this happens although the porphyrin component in **8:7₂** is exclusively excited, which was corroborated by excitation spectra of the complexes (Fig. 8.19). The latter turned out to be an exact match of the ground-state absorption of the porphyrin with maxima at 430, 560 and 600 nm. In the light of the above mentioned energy transfer processes, such findings lead to the hypothesis that upon photoexcitation the singlet excited state of the porphyrin (1.98 eV) is likely to be deactivated by transduction of the excited state energy to the fullerene (1.79 eV) in the **8:7₂** complex.

Conclusive evidence for the nature of the energy-transfer processes came from transient absorption measurements with **8** in the presence of variable concentrations of **7**. We have followed the time evolution of the characteristic singlet excited state features of **8** upon 420 nm excitation to identify the spectral features of the resulting photoproducts and to determine the absolute rate constants for the underlying deactivation processes. As reference, the transient absorption features

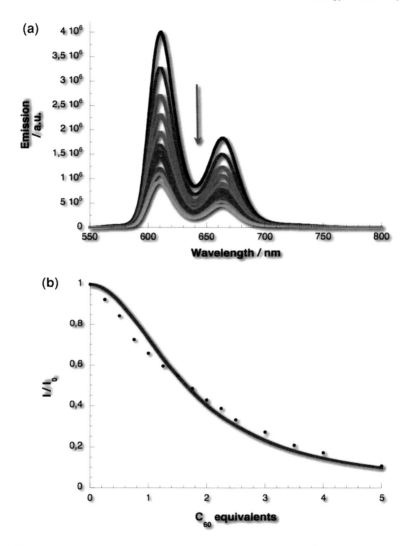

Fig. 8.18 a Room temperature fluorescence spectra of **8** (2.5×10^{-5} M) in the presence of variable concentrations of **7** (i.e. 0, 1.25×10^{-5}, 2.5×10^{-5}, 3.75×10^{-5}, 5.0×10^{-5}, 6.25×10^{-5}, 7.5×10^{-5}, 8.75×10^{-5}, 1.0×10^{-4}, 1.125×10^{-4}, and 1.25×10^{-4} M) upon 410 nm excitation, after correction for competitive ground absorption. **b** I/I_0 versus **7** relationship used to determine the association constant

of **8** have to be analyzed in order to understand the processes occurring after complexation with **7**. The differential absorption spectrum of **8** (Fig. 8.20) reveals an instantaneous formation of the singlet excited state with maxima at 455 and 800 nm and minima at 560 and 603 nm. The triplet excited-state features, on the other hand, include maxima at 500, 582, and 640 nm. Furthermore, an isosbestic

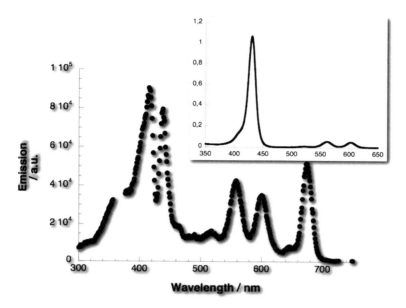

Fig. 8.19 Excitation spectrum of the **8**:**7**$_2$ complex in toluene monitoring the emission at 740 nm. *Inset*—corresponding absorption spectrum of **8**:**7**$_2$ in toluene

point at 480 nm reflects the interplay between both excited states. Additionally, the decay of the singlet excited state resembles the formation of the triplet excited state kinetics.

When turning to the transient absorption measurements of the complex **8**:**7**$_2$ (Fig. 8.21), the singlet–singlet absorption of **8** with their maxima at 455 and 800 nm as well as minima at 560 and 603 nm are seen. That confirms the successful formation of the singlet excited state of **8** despite the presence of **7**. Nevertheless, the intersystem crossing from the singlet excited state to the triplet excited state of **8** occurs on a much faster time-scale in the **8**:**7** complex resulting in the development of a new transient feature centering at 900 nm. This bears no resemblance with the triplet excited state features found in the reference spectrum of **8**. Unambiguously, the maximum at 900 nm can be attributed to the singlet excited state of **7** [8]. Moreover, the time-absorption profiles of the decay of the porphyrin singlet excited state and the formation of the fullerene features perfectly match each other, which supports the formation of one transient species (i.e. fullerene singlet excited state) at the expense of the other transient species (i.e. porphyrin singlet excited state). The lifetime of the newly formed fullerene singlet excited-state was determined as 1.3 ns identical to what has been derived for the intersystem crossing in **7** to afford the triplet manifold ($E_{Triplet} = 1.5$ eV). Finally, a maximum at 720 nm, which is a known fingerprint for the triplet excited state of C_{60} [8] evolves at the end of the decay corroborating the proposed energy transfer mechanism.

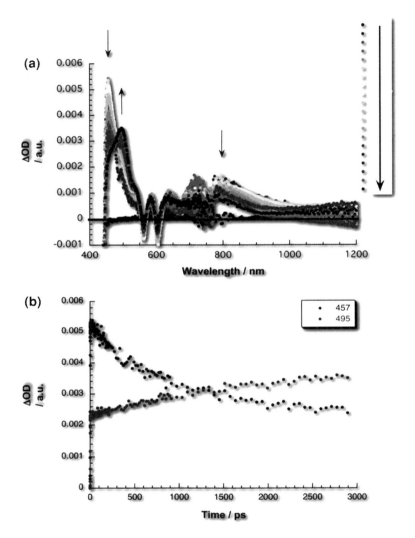

Fig. 8.20 a Differential absorption spectra (visible and near-infrared) obtained upon femtosecond flash photolysis (420 nm) of **8** in toluene with several time delays between 0 and 3000 ps at room temperature; see color code for details. *Arrows* illustrate the changes. **b** Time–absorption profile of the spectra shown above at 457 nm (*black spectrum*) and 495 nm (*red spectrum*), reflecting the singlet excited-state decay and triplet excited-state formation, respectively

The fullerene triplet quantum yields in **8**:**7**$_2$ were determined by the quantitative conversion of the triplet excited states of the porphyrin (0.78) and fullerene (i.e. 0.98) fragments into singlet oxygen. By analyzing the singlet oxygen emission at 1275 nm (see Fig. 8.22), we derived a quantum yield of 0.84.

8.2 Tunable Excited State Deactivation

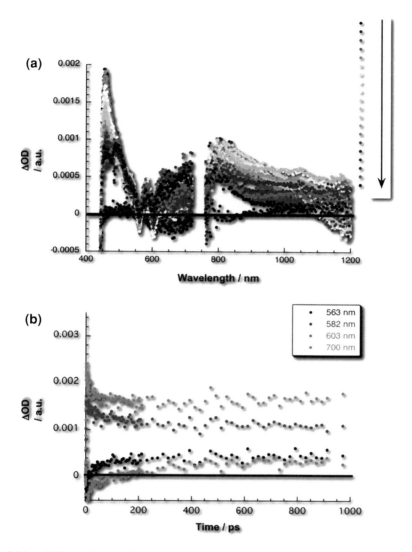

Fig. 8.21 **a** Differential absorption spectra (visible and near-infrared) obtained upon femtosecond flash photolysis (420 nm) of **8:7$_2$** in toluene with several time delays between 0 and 3000 ps at room temperature; see color code for details. **b** Time–absorption profile of the spectra shown above at 563, 582, 603, and 700 nm, reflecting the energy transfer process

Thus, it has been shown that the porphyrin chromophores act as an antenna system for transmitting its excited energy to the noncovalently associated fullerene moieties. Furthermore, the examination of excited-state characteristics revealed significant energy transfer properties of these complexes upon photoexcitation.

Fig. 8.22 Singlet oxygen emission, measured at 1275 nm, for **7** (*red spectrum*), **8** (*black spectrum*), and **8**:**7**$_2$ (*orange spectrum*) in toluene upon 355 nm excitation

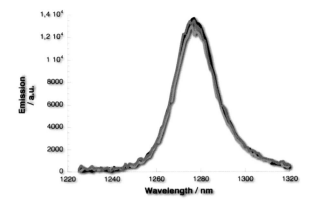

Particularly interesting is that the mediation of singlet excited-state energy is driven by *noncovalent* hydrogen-bonding interactions.

References

1. Atienza C, Insuasty B, Seoane C, Martín N, Ramey J, Guldi DM (2005) J Mater Chem 15:124
2. Pol Cvd, Bryce MR, Wielopolski M, Atienza-Castellanos C, Guldi DM, Filippone S, Martín N (2007) J Org Chem 72:6662
3. Goldsmith RH, Sinks LE, Kelley RF, Betzen LJ, Liu W, Weiss EA, Ratner MA, Wasielewski MR (2005) Proc Natl Acad Sci U S A 102:3540
4. Dewar MJS, Zoebisch E, Healy EF, Stewart JJP (1985) J Am Chem Soc 107:3902
5. Chang S-K, Hamilton AD (1988) J Am Chem Soc 110:1318
6. Gnichwitz J-F, Wielopolski M, Hartnagel K, Hartnagel U, Guldi DM, Hirsch A (2008) J Am Chem Soc 130:8491
7. Hager K, Hartnagel U, Hirsch A (2007) Eur J Org Chem 12:1942
8. Guldi DM, Prato M (2000) Acc Chem Res 33:695

Chapter 9
Electron Transfer Systems

9.1 p-Phenyleneethynylene Molecular Wires

We have seen in Chaps. 5 and 6 that π-conjugated oligomers are the most promising candidates for molecular wires. They provide efficient electronic coupling between two electroactive units, i.e. donor and acceptor. Among the variety of π-conjugated systems, *oligo*(*para*-phenylenevinylene)s (*o*PPVs) have emerged as particularly interesting in terms of their molecular-wire behavior due to their outstanding conduction properties. Above, it has been demonstrated that when connecting *o*PPV-bridges of different length to an electron-accepting C_{60} and an electron-donating *ex*TTF moiety, the *o*PPV moieties promote wirelike behavior for donor–acceptor distances of 40 Å and beyond. The *ex*TTF–*o*PPV$_n$–C_{60} conjugates met all the requirements rendering it an ideal molecular wire:

- energetic matching between the donor (acceptor) and bridge levels,
- sufficient electronic coupling between the donor and acceptor units induced by the bridge orbitals,
- small attenuation factor β, namely $\beta = 0.01 \pm 0.005$ Å^{-1}.

Further, the investigation of the conduction behavior as a function of distance revealed nearly distant-independent charge-transfer characteristics [1]. These findings were corroborated by probing analogous systems containing porphyrins as electron donors instead of *ex*TTF [2, 3].

To this end, intramolecular electron transfer along conjugated *o*PPV chains has been tested in several other donor–acceptor conjugates employing a variety of electron donors—anilines [4], porphyrins [5] and ferrocenes [6]—in combination with fullerenes as electron acceptors. The quintessence of this study is that energy matching between the donor and bridge components is key for achieving molecular-wire behavior. Quantum-chemical calculations showed a competition between a direct superexchange process and a two-step "bridge-mediated" process, whose efficiency depends primarily on the length and nature of the conjugated bridge [7].

Encouraged by these results, we shifted our focus to very similar systems, namely *oligo(para*-phenyleneethynylene)s (*o*PPEs). When compared to *o*PPVs, replacing the double bonds by triple bonds adds additional rigidity to the π-system. To elaborate the influence of such a structural modification especially on the electron-transfer properties of the *o*PPEs, we have tested a series of novel donor–acceptor conjugates, i.e. *ex*TTF–*o*PPE$_n$–C$_{60}$ and ZnP/H$_2$P–*o*PPE$_n$–C$_{60}$. As another extension *meta*-connected phenyleneethynylene bridges (*o*MPEs) were investigated to elaborate the influence of connectivity on the π-conjugation. The systematic study of electronic *para*- versus *meta*-properties by means of photophysical and quantum chemical methods will be described in the following chapters.

9.1.1 ex*TTF*–o*PPE*–C$_{60}$ Donor–Acceptor Conjugates

The group of Professor Nazario Martín developed the synthesis of several donor–acceptor arrays containing π-conjugated *o*PPEs of different length linking π-extended tetrathiafulvalene (*ex*TTF) as electron donor and C$_{60}$ as electron acceptor. Thereby, they systematically increased the length of the molecular wire from the monomer to the trimer. A detailed description of the corresponding synthesis is described elsewhere [8].

The structures of *ex*TTF–*o*PPE$_n$–C$_{60}$ (**9b–d**) and the reference compounds used throughout the electrochemical and photophysical studies are represented in Fig. 9.1.

9.1.1.1 Electrochemistry

Prior to any photophysical experiments, cyclic voltammetry experiments were performed with *ex*TTF–*o*PPE$_n$–C$_{60}$ **9b–9d** and the reference compounds *o*PPE$_n$–C$_{60}$ **10**, **11** to provide a vital impression of the energy-levels in these compounds.

For **10** and **11** four quasi-reversible reduction waves were found, which are attributed to the reduction of the C$_{60}$ moieties leading correspondingly to the mono-, di-, tri- and tetraanion. Nevertheless, the reduction potentials were all shifted toward more negative values relative to pristine C$_{60}$. Saturation of a double bond in the fullerene skeleton raises the LUMO energy in **10** and **11**. Furthermore, two oxidation waves stemming from oligomer centered processes were found.

The cyclic voltammetry studies of **9b–d** also revealed the presence of four quasi-reversible reduction waves at values similar to those found for the reference compounds (**10**, **11**). Importantly, when comparing the oxidation potentials of **10** and **11** with those of the triads **9b–d** and *ex*TTF, the potentials converge with those of pristine *ex*TTF (around 0.33 V). In the triads, the first oxidation is found anodically shifted relative to pristine *ex*TTF. This confirms its strong donating character. On the other hand, with increasing bridge length the *ex*TTF oxidation

9.1 p-Phenyleneethynylene Molecular Wires

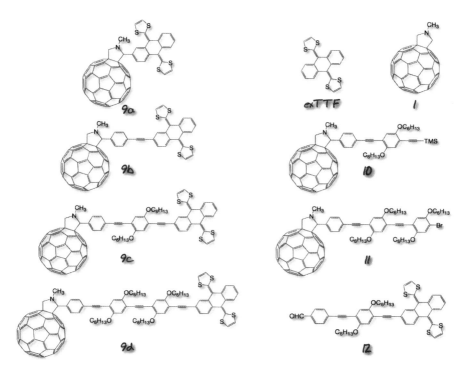

Fig. 9.1 Schematic representation of the $exTTF$–$oPPE_n$–C_{60} triads and their appropriate reference compounds

remains at nearly constant potential. One additional reduction potential wave could be found in **9b–d**. Interestingly, this process is attributed to the reduction of the oligomer units.

In general, the redox potentials of $exTTF$–$oPPE_n$–C_{60} (**9b–d**) match the redox potentials of the pristine donors and acceptors, attesting a lack of significant electronic interactions between the building blocks in the ground state. Similar assumptions are conducted from inspecting the corresponding absorption spectra (i.e. references and conjugates).

9.1.1.2 Photophysics

Figure 9.2 shows a representative example of the UV/vis absorption characteristics of $exTTF$–$oPPE_n$–C_{60} in comparison with their reference compounds **10** and **12**. The spectra reveal the typical features of the building blocks, i.e. C_{60}, $exTTF$ and $oPPE$. Clearly visible is a bathochromic shift of **9c** relative to the corresponding reference compound **10**. Implicit is an extended π-conjugation in the triads. Likewise, when inspecting the different oligomers, namely **9b–d**, their absorption maxima gradually shift to the red with increasing number of repeat units (not

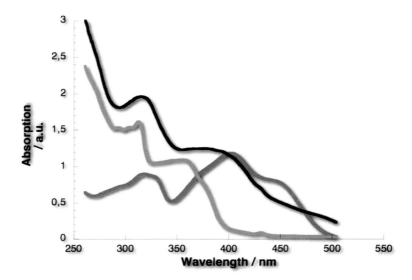

Fig. 9.2 UV–visible absorption spectra in toluene of **9c** (*black line*), **10** (*orange line*) and **12** (*red line*)

shown). Again, this is due to the extension of the π-system. The spectral signatures of the building blocks in the triads reflect the lack of electronic interactions between them in the ground state corroborating the findings stemming from cyclic voltammetry experiments.

In summary, the ground-state features of $exTTF-oPPE_n-C_{60}$ prove the electron donor/acceptor character in these architectures and we should move on to the characterization of the excited state features, which revealed remarkable charge-transfer processes in these systems.

First, the steady-state emission spectra are evaluated. Inspecting the fluorescence of $C_{60}-oPPE$ discloses the strong emission characteristics of $oPPEs$, which was already described in Sect. 8.1.1. All $oPPE$ references fluoresce strongly throughout the visible region. Thus, studying excited-state interactions with C_{60} in $C_{60}-oPPE$ and $exTTF-oPPE_n-C_{60}$ was straightforward. In addition to the characterization of the charge transfer processes in $exTTF-oPPE_n-C_{60}$, we added the photophysical characterization for the 0-mer (**9a**), i.e. an electron donor system in which $exTTF$ is directly linked to N-methylpyrolidine group of C_{60} (see Fig. 9.1).

Since the photophysical assays of the $oPPE$ reference compounds and the $C_{60}-oPPEs$ have already been evaluated in Sect. 8.1.1, only the most important features will be mentioned in order to understand the processes in photoexcited $exTTF-oPPE_n-C_{60}$.

As already mentioned, the $oPPEs$ fluoresce strongly in the visible range with quantum yields close to unity and a length-dependent red-shift of the emission maxima reaching from 400 to 490 nm for the monomer and the trimer, respectively. In the $C_{60}-oPPEs$ this $oPPE$ fluorescence is remarkably quenched due to energy transfer to the energy accepting C_{60}. Thus, in the presence of C_{60} the excited state instantaneously deactivates notwithstanding the fact that the oPPE fluorescence pattern is still preserved. Likewise, in the near-infrared region the spectra reveal the fluorescence features of C_{60} (**1**). Consequently, these findings corroborate the singlet excited-state energy transfer from the oPPE to C_{60} upon 355 nm photoexcitation, which directs the light almost quantitatively to the oPPE oligomer and not to C_{60}. When comparing the fluorescence quantum yields in toluene solutions of C_{60}–oPPE with that of **1**, a rapid intramolecular transduction of energy starting with photoexcited oPPE and leading to the singlet excited state of C_{60} ($^1{}^\star C_{60}$) was quantified. Excitation of the fullerene, on the other hand, leads directly to $^1{}^\star C_{60}$

Inspecting the fluorescence in exTTF–oPPE$_n$–C_{60} (Fig. 9.3), unveiled an appreciable quenching of the fullerene emission which maximizes at 710 nm. The quantum yield of the fullerene fluorescence in **9b** amounts, for instance, in THF to 0.55×10^{-4}. This is more than one order of magnitude lower in comparison with **1** (6.0×10^{-4}). Furthermore, the quantum yields depend on the length of the oligomer bridge:

- 0-mer **9a**: 0.18×10^{-4},
- monomer **9b**: 0.55×10^{-4},
- dimer **9c**: 1.8×10^{-4},
- trimer **9d**: 6.0×10^{-4} (i.e., no notable interactions were found).

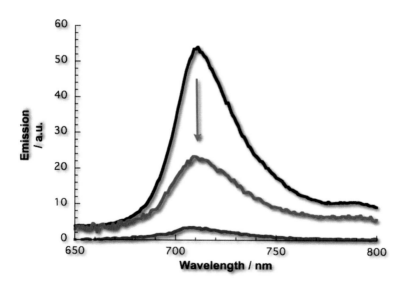

Fig. 9.3 Quenching of the C_{60} emission in exTTF–oPPE$_n$–C_{60} **9b**, **9c** upon excitation at 355 nm with matching optical density at 355 nm in THF: C_{60}-reference **1** (*black*), monomer **9b** (*dark blue*) and dimer **9c** (*light blue*)

By means of time-resolved fluorescence studies we were able to determine the C_{60} fluorescence deactivation rates, as 2.1×10^{10} s^{-1} in **9a**, 6.6×10^{9} s^{-1} in **9b** and 1.3×10^{9} s^{-1} in **9c**. Importantly, the indulging trend resembles the relationship between the quantum yields of the conjugates (**9a–d**) and reference (**1**). In short, an intensified excited-state deactivation emerges with decreasing bridge length. However, no measurable decay rates were found for the trimer **9d**. Conclusively, the indirect or direct population of $^{1\star}C_{60}$ possibly leads to an exothermic electron-transfer reaction, resulting in the radical-ion-pair state: $exTTF^{\bullet+}$–$oPPE_n$–$C_{60}^{\bullet-}$.

Evidence for the formation of the radical ion pair, was found in transient absorption spectroscopy measurements. In Sect. 8.1.1 the results of transient absorption studies on the oPPEs and C_{60}–oPPEs have already been introduced and will now be exemplified by appropriate spectra.

Comparative to the outcome obtained with the C_{60}–oFL–C_{60} systems (Sect. 8.1.2), the spectral features of the oPPEs reveal the nearly instantaneous generation of metastable singlet excited-state transients. Typical characteristics involve ground-state bleaching in the 400–450 nm range and new transient absorptions between 600 and 1200 nm (Fig. 9.4). The product of the photoexcitation is the singlet excited state of the oligomer.

Like in the case of C_{60}–oFL, the presence of C_{60} in C_{60}–oPPE accelerates the singlet excited-state deactivation of the oPPE moieties and leads to the development of new transients. Despite the presence of C_{60}, the oPPE typical transient bleaching (below 450 nm) and transient maxima (around 700 nm) are still present. In close agreement with the ground-state absorption spectra, a 387 nm excitation exclusively excites the oPPE units. In contrast to the oPPE references, the product of the singlet excited-state decay is not the oPPE triplet anymore. Decay rates on the order of 10^{12} s^{-1}, which mirror image the quantitative fluorescence quenching of the oligomer emission, dominate the singlet excited-state deactivation. As a result, the very fast formation of the C_{60} singlet excited state is inferred from the new transient maximum at 880 nm (Fig. 9.5). Following its formation, the singlet excited-state of C_{60} decays to afford the corresponding triplet manifold. A transient maximum at 700 nm evolves toward the end of the femtosecond experiment (i.e. after 3.0 ns) and is an unambiguous indication for the triplet excited state. Kinetically, the singlet excited-state decay and the triplet excited state growth are linked. The corresponding intersystem crossing rates were determined for C_{60}–oPPE as 6.5×10^{8} s^{-1}. Complementary transient absorption experiments on the nanosecond timescale with C_{60}–oPPE revealed the same triplet transient at 700 nm, which decays with multi-exponential kinetics. Overall, our observations corroborate the successful transduction of singlet excited-state energy to the C_{60} core.

Yet, the spectroscopic features of the transients change again in the presence of exTTF. The instantaneous grow-in of the singlet excited-state state absorption of C_{60} at 880 nm appears, once more affirming the successful excitation of C_{60}. In contrast to C_{60}–oPPE, the decay of the singlet–singlet absorption is fast without giving rise to any intersystem crossing, which would generate the corresponding

9.1 *p*-Phenyleneethynylene Molecular Wires

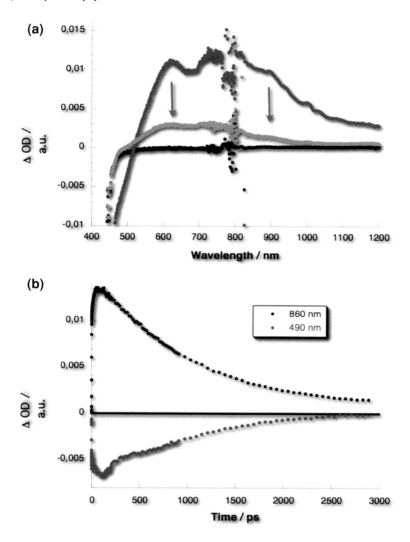

Fig. 9.4 a Differential absorption spectrum (visible and near-infrared) obtained upon femtosecond flash photolysis (420 nm) of solutions of reference trimer *o*PPE in nitrogen-saturated THF with time delays between 0 and 3000 ps at room temperature (*black* = 0 ps, *red* = 1 ps, and *orange* = 2900 ps). **b** Time–absorption profile of the spectra shown above at 490 and 860 nm to monitor the formation and decay of the singlet excited state of trimer *o*PPE

$^3\star C_{60}$. Instead, the spectral signatures of the one-electron oxidized *ex*TTF$^{\bullet+}$ radical cation at 660 nm and of the one-electron reduced $C_{60}^{\bullet-}$ radical anion at 1010 nm were detected (Fig. 9.6). The singlet excited-state lifetimes, as determined from an average of first-order fits from time–absorption profiles at various wavelengths are listed in Table 9.1.

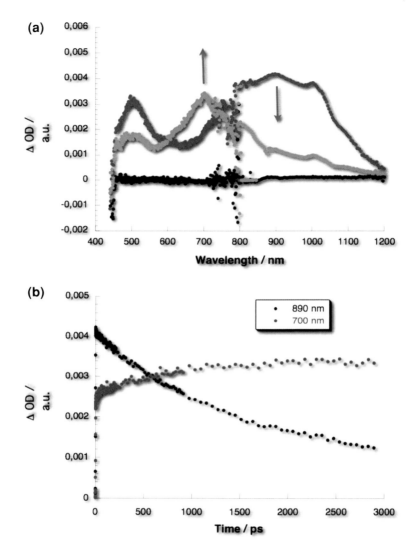

Fig. 9.5 a Differential absorption spectrum (visible and near-infrared) obtained upon femtosecond flash photolysis (420 nm) of solutions of reference C_{60}–oPPE **10** in nitrogen-saturated THF with time delays between 0 and 3000 ps at room temperature (*black* = 0 ps, *red* = 1 ps, and *orange* = 2900 ps). **b** Time–absorption profiles of the spectra shown above at 700 and 890 nm to monitor the intersystem crossing between the singlet and triplet excited states of C_{60}

Conclusively, in the presence of a strong electron donor, namely *ex*TTF, the resulting product of the photoexcitation is not the triplet excited state of C_{60} but the energetically lower-lying radical-ion-pair state, $exTTF^{\bullet+}$–$oPPE_n$–$C_{60}^{\bullet-}$. Here, it should be stressed that the spectral identification of the radical ion pair holds for

9.1 p-Phenyleneethynylene Molecular Wires

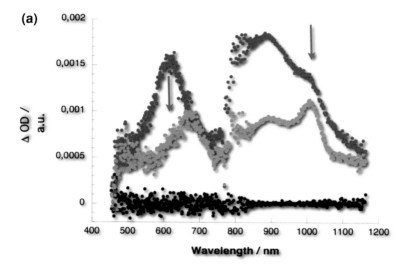

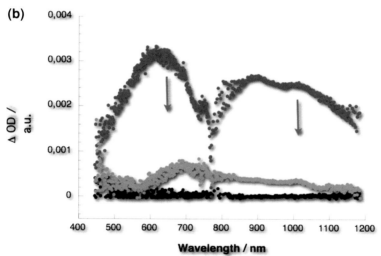

Fig. 9.6 **a** Differential absorption spectrum (visible and near-infrared) obtained upon femtosecond flash photolysis (477 nm) of solutions of monomer **9b** in nitrogen-saturated THF with time delays between 0 and 3000 ps at room temperature (*black* = 0 ps, *red* = 1 ps, and *orange* = 2900 ps). **b** Differential absorption spectrum (visible and near-infrared) obtained upon femtosecond flash photolysis (477 nm) of solutions of dimer **9c** in nitrogen-saturated THF with several time delays between 0 and 3000 ps at room temperature (*black* = 0 ps, *red* = 1 ps, and *orange* = 2900 ps)

the 0-mer **9a**, monomer **9b**, and dimer **9c**. In the trimer **9d**, a very broad transient dominates the region of interest. Formation of the radical ion pair is, however, ruled out. Specific details on that matter will follow later. Both radical-ion species, i.e. $ex\text{TTF}^{\bullet+}$ and $C_{60}^{\bullet-}$, decay with similar rates to the corresponding singlet ground

Table 9.1 Photophysical properties of the *ex*TTF–*o*PPE–C_{60} triads **9a–d** and the *N*-methylfulleropyrolidine **1**

	Φ_{fl}	τ_1^a [ns]	k_{CS} [s^{-1}]	k_{CR} [s^{-1}]
C_{60} reference	6.0×10^{-4}	1.233	–	–
0-mer **9a**	0.18×10^{-4}	–	2.1×10^{10}	4.9×10^6 s^{-1}
monomer **9b**	0.55×10^{-4}	–	6.6×10^9	1.1×10^6 s^{-1}
dimer **9c**	1.8×10^{-4}	0.380	1.3×10^9	3.8×10^5 s^{-1}
trimer **9d**	–	(0.793)	3.9×10^8	not detectable

[a] Singlet lifetime

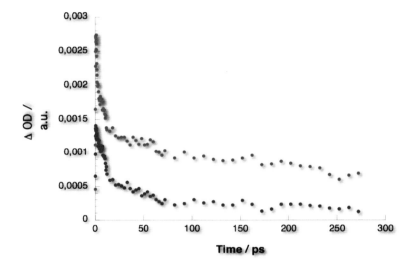

Fig. 9.7 Time–absorption profile of the spectra shown above at 1010 nm to monitor the formation of the radical-ion-pair state (*blue* = monomer **9b**, *light blue* = dimer **9c**)

states (**9a**: 4.9×10^6 s^{-1}, **9b**: 1.1×10^6 s^{-1}, **9c**: 3.8×10^5 s^{-1}, Fig. 9.7) driven by an exothermic ($-\Delta G = 1.0$ eV) energy gap.

Finally, relating the charge-separation and charge-recombination dynamics with the distance between electron donor and acceptor (i.e. center-to-center distance R_{CC}), allows the evaluation of the attenuation factor β. The corresponding relationships in THF are represented in Fig. 9.8. Both reveal linear dependencies and the derived attenuation factors are in perfect agreement with each other:

- charge separation: $\beta = 0.21 \pm 0.05$ Å^{-1},
- charge recombination: $\beta = 0.20 \pm 0.05$ Å^{-1}

A comparison with the reported β values of the *o*PPV bridges ($\beta = 0.01 \pm 0.05$ Å^{-1}) [1] infers a significantly poorer molecular-wire character for *o*PPEs, since the β factors differ by a factor of 20.

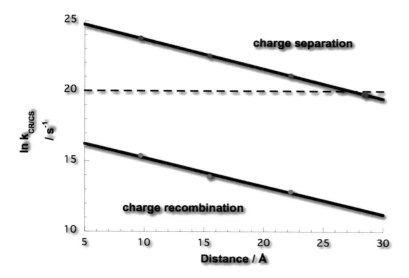

Fig. 9.8 Centre-to-centre distances (R_{CC}) dependence of charge-separation (ln k_{CS}) and charge-recombination (ln k_{CR}) rate constants in exTTF–oPPE–C_{60} in nitrogen-saturated THF at room temperature. The slope represents β. The *dashed line* represents the singlet lifetime of C_{60}. In *blue*, the extrapolated charge-separation rate for the trimer **9d** is represented

A possible explanation for the lack of electron-transfer characteristics in the trimer **9d** is derived when extrapolating the linear relationship in Fig. 9.8 to the distance of the trimer. As a matter of fact, the charge-separation would not be able to compete with the intrinsic singlet lifetime of C_{60} (i.e. dashed line). This, in turn, explains the lack of fullerene emission quenching in **9d**. Nevertheless, the photophysical assays clearly established that oPPE bridges effectively mediate electron-transfer processes over distances up to 20 Å. These findings were further corroborated by quantum mechanical calculations.

9.1.1.3 Molecular Modeling

First, the molecular geometries should be considered, since the relative arrangement of the phenyl rings in the oPPE bridge is crucial for the preservation of the π-conjugation in the investigated exTTF–oPPE$_n$–C_{60} triads. For this reason, we have employed a number of DFT and semiempirical calculations.

Semiempirical calculations suggested that in the minimum-energy structure the dihedral angle formed by the phenyl rings approaches zero. These results were confirmed by using DFT calculations at the *B3LYP*/6−31*G** level and crosschecked by using *B3PW*91/6−31*G** calculations. The deviation from planarity turned out to be less than ± 12°. Furthermore, single-point calculations at the *B3LYP*/6−311*G*** and the *B3PW*91/6−311*G*** levels reveal a very low

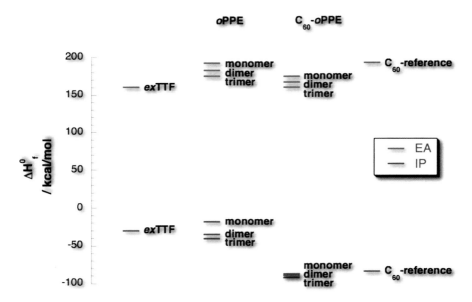

Fig. 9.9 Calculated *AM1** ionization potentials IP (*red*) and electron affinities EA (*blue*) of the pristine building blocks of the triads

rotation barrier of the phenyl rings in the π-conjugated bridge (i.e. less than 2.0 kcal/mol). Rotations of the phenyl unit adjacent to the pyrrolidine and the benzene moiety of exTTF connected to the oligomer are similarly not restricted by the relatively low rotation barrier of less than 1.0 kcal/mol.[1]

On inspecting the electronic structure and energy-levels of the DBA triads, further evidence for the described electron-transfer characteristics was discovered. Figure 9.9, for instance, shows the calculated (*AM1** [9]) ionization potentials (IPs) and electron affinities (EAs) of *ex*TTF, C_{60}–pyrrolidine, the pristine *o*PPE oligomers and C_{60}–*o*PPE. Thereby, *ex*TTF exhibits the lowest EA and those that matches that of the oligomer building blocks. With the addition of C_{60} to the oligomers, the IP drops to approximately the IP of C_{60}–pyrrolidine, which suggests strong electronic coupling between the fullerene and the bridge. On the other hand, the electron-accepting features of the fullerene are represented by its highest EA, which confirms the electron-transfer pathway of the systems.

Moreover, the HOMO/LUMO orbital schemes of the triads (Fig. 9.10) manifest this donor–acceptor character. The HOMO is strongly localized on *ex*TTF and reaches into the *o*PPE bridge, which facilitates electron injection into the bridge. However, in contrast to *ex*TTF–*o*PPV$_n$–C_{60}, in which the HOMO is completely

---

[1] Rotations in the monomer **9b**, dimer **9c** and trimer **9d** do not seem to influence the electronic structure and the coupling between donor and acceptor. The rotational barriers comply with these found in the *ex*TTF–oPPV$_n$–C_{60} triads (0.34 kcal/mol).

9.1 p-Phenyleneethynylene Molecular Wires

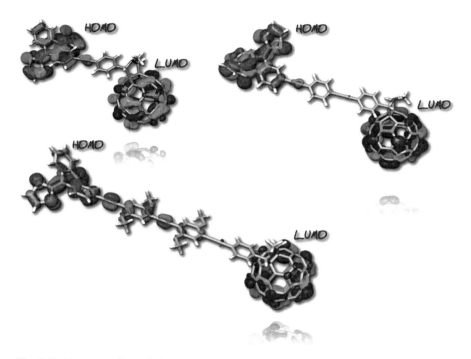

Fig. 9.10 Representation of the HOMO (*red–green*) and LUMO (*blue–green*) orbitals of *ex*TTF–*o*PPE–C_{60} as calculated at the *B3LYP*/6–31*G** DFT level

conjugated throughout the whole bridge, in the corresponding *ex*TTF–*o*PPE$_n$–C_{60} the localization is more pronounced. Hence, injection of an electron into the bridge in *ex*TTF–*o*PPV–C_{60} is favoured by the better orbital overlap between *ex*TTF and the oligomers.

Local-electron-affinity mappings of *ex*TTF–*o*PPE$_n$–C_{60} and *ex*TTF–*o*PPV$_n$–C_{60} are represented in Fig. 9.11. Significant differences between the two systems can be seen when comparing the conjugation in the bridge. In *ex*TTF–*o*PPV$_n$–C_{60}, the local electron affinity is homogeneously distributed throughout the whole bridge, whereas in the *o*PPE systems local maxima (red) can be found on the phenyl rings and minima (yellow) on the triple bonds. This is due to the polarising character of the triple bonds and their shorter bond lengths relative to the double bonds. Thus, the electron-transfer pathway through the *o*PPE bridge is interrupted by these ethynylene linkers. This strongly influences the charge-separation process and explains the difference in electron-transfer properties between *ex*TTF–*o*PPE$_n$–C_{60} and *ex*TTF–*o*PPV$_n$–C_{60}.

In summary, the electronic structure of all triads confirms the donor–bridge–acceptor character of the *ex*TTF–*o*PPE$_n$–C_{60} systems and suggests that the HOMO → LUMO transition would represent a nearly complete charge-transfer excitation with a very low extinction coefficient. Since these findings are based on

Fig. 9.11 Local-electron-affinity maps of $exTTF-oPPV_3-C_{60}$ (*top*) and $exTTF-oPPE_3-C_{60}$ (*bottom*) as viewed with Tramp 1.1d

Table 9.2 Excited-state properties predicted by quantum chemical calculations for compound **9d** (top: CIS including six active orbitals; bottom: CIS including ten active orbitals)

State	λ_{exc} [nm]	f	$\Delta\mu$ [D]	Character	Involved configuration
CIS = 6					
S1	395	0.000	135.3	CT	HOMO → LUMO
S2	385	0.000	136.9	CT	HOMO → LUMO+1
CIS = 10					
S2	376	0.066	3.6	LE1	HOMO-4 → LUMO
S6	285	0.554	4.0	LE2	HOMO-4 → LUMO+3
S7	278	0.000	129.3	CT	HOMO → LUMO, HOMO-2 → LUMO
S8	277	0.001	69.8	BCT	HOMO-2 → LUMO, HOMO-3 → LUMO

ground-state calculations and the photophysical processes occur on the excited-state potential surface, we have carried out further calculations in order to determine excited-state properties.

To elucidate the geometrical relaxation processes postulated on the basis of the existence of very short-lived components in the time-resolved spectroscopic techniques described above, we optimized the minimum-energy geometries of the excited-states in vacuum using configuration interaction (CI) methods. Singles-only configuration interaction (CIS) calculations predict that the HOMO → LUMO transition makes a major contribution to the charge-transfer (CT) state and causes a large change in dipole moment (Table 9.2). One-electron excitation from the HOMO to the LUMO contributes to the CT state with zero oscillator strength (f) (Table 9.2). In addition, in **9d** two symmetric bridge charge-transfer states (BCT) could be found close in energy to the CT states with the lower change in dipole moment (69.8 Debye). These states correspond to the local-electron-affinity maxima centred on the oPPE rings, which underlines the utility of the local electron affinity at the surface as a fast scanning method to elucidate possible CT states. These bridge CT states can compete with the charge transfer from exTTF to C_{60}. The excitation wavelengths for charge-transfer excitations (in solution) were calculated to $\lambda_{exc} = 350$ nm with a redshift that is dependent on the length of the oligomer but independent of solvent polarity. Similar behavior was found for the fluorescence maxima of **9b–d** with a solvent-dependent fluorescence, which maximizes between $\lambda_{max} = 450$ and $\lambda_{max} = 500$ nm and also depends on the length of the oligomeric bridge. However, regarding the fluorescence maximum of the BCT state of **9d** at 380 nm with a 261 nm excitation in simulated diethyl ether, an overlap between the CT fluorescence that reaches a maximum at 450 nm is possible and depends on the quantum yield.

Scrutinizing the electrostatic potential of the AM1-optimized ground states and the corresponding CT states mapped onto the molecular surface (Fig. 9.12) reveals that in all CT states of the exTTF–oPPE$_n$–C_{60} triads the positive charge is localised on exTTF (red) and the negative charge on C_{60} (blue). Alternatively, in the BCT

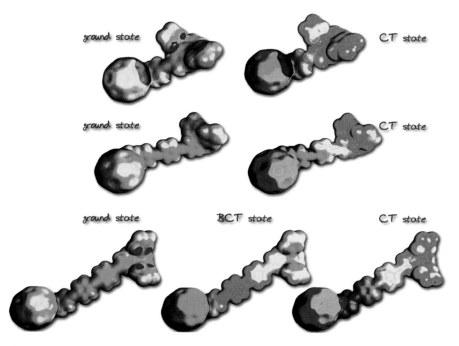

Fig. 9.12 Electrostatic potential as calculated by AM1 CIS for the ground and charge-separated states of **9b–d**. The positive charge is represented in *red* and the negative in *blue*

state the charge is localised on one of the phenyl rings. The corresponding symmetric BCT state was also found in the CI calculations.

Further examination of the excited states reveals a strong dependence of their energy upon solvent polarity that resembles the experimental trends. Single-point calculations on the relaxed structures of the ground and excited states in the simulated solvents hexane ($\epsilon = 2.023$), CCl$_4$ ($\epsilon = 2.229$), benzene ($\epsilon = 2.274$), ether ($\epsilon = 4.197$), chloroform ($\epsilon = 4.806$), methylene chloride ($\epsilon = 8.930$), pyridine ($\epsilon = 12.40$), acetone ($\epsilon = 20.56$), ethanol ($\epsilon = 24.55$), nitrobenzene ($\epsilon = 43.82$), acetonitrile ($\epsilon = 35.94$) and dimethyl sulfoxide ($\epsilon = 46.45$) were performed in order to judge the solvent effects on the relative stabilities of the different states for **9b–d**.

Figure 9.13 shows the dependence of the calculated heats of formation, ΔH_f, of the discussed states of **9d** on solvent permittivity, i.e. $\frac{\epsilon-1}{2\epsilon+1}$. By optimizing the local-excited (LE) state, the BCT state and the CT state of **9d**, it was possible to determine the energy levels of the three different Franck–Condon states and gain extensive insight into the electron-transfer pathway. The calculations reveal that the relative energies of the different states and conformations including their solvent dependence are reproduced remarkably well according to the experimental trends. Figure 9.13 suggests that with higher solvent polarity, the energy splitting between the different states increases significantly, whereas their relative energies decrease due to solvent stabilization. Equally important, one can see that the BCT becomes more and more accessible as the polarity of the solvent increases. This behavior was also observed in the photophysical experiments. However, no crossing between the CT and BCT state, even in highly polar media, was observed during the calculations.

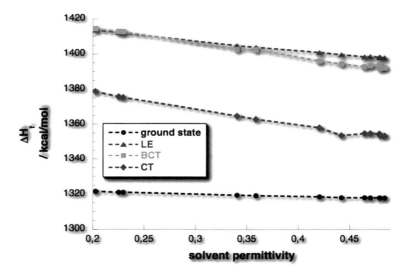

Fig. 9.13 Dependence of the calculated heats of formation, ΔH_f, for the discussed states of **9d** on solvent permittivity: ground state (*black*), local excited state (*blue*), BCT (*green*), CT (*red*)

The results for the monomer **9b** and the dimer **9c** reveal equal solvent dependence of the discussed states. However, in these two compounds no BCT is present. The heat of formation, ΔH_f, of the CT states is increasing almost linearly from monomer to trimer. These trends prove the observations resulting from the photophysical measurements and confirm the suggested charge-transfer behavior of exTTF–oPPE$_n$–C_{60}. Thus, the calculations support the hypothesis that in all solvents an excitation of the triad results in the CT state, which substantiates our interpretation of the electron-transfer mechanism.

Furthermore, we have calculated the Coulson charge on exTTF and C_{60} in order to confirm one-electron transfer from HOMO to LUMO. The values were determined to $1.0e^-$ and $-1.1e^-$ for exTTF and C_{60}, respectively. These findings indicate complete transfer of one electron from the donor to the acceptor with some delocalization of the positive charge (4.6%) into the bridge. exTTF is clearly oxidized with the positive charge localized on the sulfur atoms ($0.6e^-$ on each), whereas C_{60} is reduced with the charge being delocalized over the entire carbon cage.

Finally, we have used the optimized excited-state structures of the three different states, i.e. local excited, BCT and CT, to calculate the reaction coordinate in photoexcited **9d**. The energy diagrams were determined from the two low-energy conformations of each state, namely the optimized ground-state geometry and the optimized excited-state geometry, by performing a *linear synchronous transit* (LST) calculation upon changing the electronic configuration from the ground-state geometry to the excited-state geometry of the given state [10]. Figure 9.14 represents the calculated charge-transfer pathway in **9d**. Following the reaction coordinate of the locally excited state (top left), we pass several points at which the curves approach each other. At these points, the energy difference between two separate states is sufficiently low for these states to cross, i.e. the molecule changes its electronic configuration from one state to the other. The electrostatic potential (neutral = green, positive = red, Negative = blue) of each state is mapped onto the molecular surface and describes the changes of electronic structure upon changing the state. Thus, following the highest energy gradient, we can establish a possible reaction path (blue dotted line) and examine the states that have been passed until the molecule relaxes in the minimum energy configuration of the charge-separated state (bottom right). The following reaction mechanism has been determined from these calculations: Photoexcitation leads to a locally excited state (No. 1) with no change in dipole moment. The excited-state energy is then funneled to the exTTF moiety (No. 2) which in turn induces a stepwise hole transfer involving the oPPE bridge (No. 3). At the conclusion of the hole-transfer reaction, a local excitation of exTTF (No. 4) and the final charge-separated state (No. 5) is reached in one single step (indicated by the blue arrow), due to a high energy difference between the two last steps. Importantly, these findings suggest hole transfer rather than electron transfer as the dominant mechanism for the charge-transfer processes.

In conclusion, quantum mechanical calculations disclose the donor–bridge–acceptor character of the exTTF–oPPE$_n$–C_{60} systems and suggest that the HOMO → LUMO transition would represent a nearly complete charge-transfer

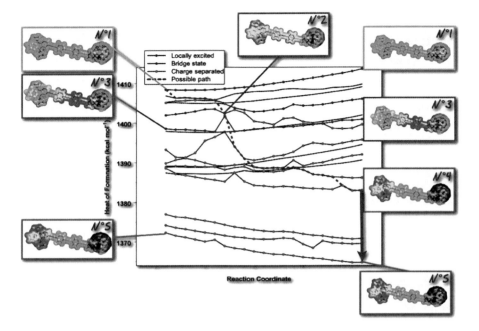

Fig. 9.14 Reaction pathway for photoexcited $ex\text{TTF}-o\text{PPE}_3-\text{C}_{60}$ obtained from linear synchronous transit calculations on the optimized geometries of excited states. A possible reaction mechanism is indicated by the *blue dotted line* and suggests a hole transfer. For further details see text

excitation with a very low extinction coefficient. Following, the LST reveals that photoexcitation induces a rapid intramolecular charge separation to generate a radical-ion-pair state.

9.1.1.4 Summary

Indisputably, photoexcitation is followed by a rapid deactivation of the singlet-excited state of the oPPE moiety resulting in the generation of a charge-separated species, i.e. the radical-ion-pair state $ex\text{TTF}^{\bullet+}-o\text{PPE}_n-\text{C}_{60}^{\bullet-}$, which is apparently lower in energy than the corresponding triplet state of C_{60}. The radical ion pairs decay on the µs time-scale with charge-recombination rates that prove wire-like behavior with attenuation factors of $\beta = 0.20 \pm 0.05 \text{ Å}^{-1}$. A schematic representation of the possible reaction pathways is given in the energy diagram of Fig. 9.15.

9.1.2 $H_2P/ZnP-oPPE-C_{60}$ Donor–Acceptor Conjugates

Section 9.1.1 scavenged the electron-transfer processes along *oligo*-(p-phenyleneethynylene) molecular wires as implemented in $ex\text{TTF}-o\text{PPE}_n-\text{C}_{60}$ donor–acceptor

9.1 p-Phenyleneethynylene Molecular Wires

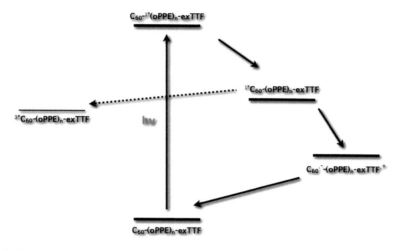

Fig. 9.15 Schematic illustration of the reaction pathways in photoexcited **9b–d**

conjugates. The attenuation factor β was established as $0.20 \pm 0.05 \, \text{Å}^{-1}$, which stands for fairly good electron-transfer characteristics. Consequently, the next challenge was to determine whether this β value is specific for oPPEs or whether it describes the donor–acceptor system as a whole. In particular, we were interested to see if this β-value depends on the donor and acceptor implemented. In that context, together with the group of Prof. Tagliatesta from the University of Rome we have developed novel donor–acceptor systems based on C_{60} as electron acceptor and oPPE molecular bridges. In variance to the previous systems, a free base porphyrin and its zinc complex served as donor moiety [11]. To enhance the electronic interaction through the extended π-system, the molecular bridges have been directly linked to the β-pyrrole position of the porphyrin ring. New examples of DBA systems, where, for the first time, the *meso*-phenyl ring of the macrocycle is not utilized to link the porphyrin and fullerene moieties were obtained. As a consequence, altering the chemical and physical properties of the tetrapyrrole ring is facilitated.

Similarly to the exTTF–oPPE$_n$–C$_{60}$ conjugates, a thorough investigation of photoinduced charge-transfer processes in H$_2$P/ZnP–oPPE$_n$–C$_{60}$ (**15a–c**) by electrochemical, spectroscopic and quantum mechanical methods will be presented in the next part of this thesis. The synthesis is described elsewhere [11].

The structures of the newly synthesized systems are represented in Fig. 9.16. Notably, the β-pyrrole position is the junction between the bridge and the porphyrin. Importantly, the carbon–carbon triple bond directly attached to the pyrrole ring affords an extended π-conjugation, which ranges over the whole oPPE residue. Thus, the distance between the porphyrin moiety and the fullerene is completely conjugated and only interrupted by the sp^3 carbon atom of the pyrolidine ring, where the bridge has been attached (see Fig. 9.16). Hence, these structures are ideally suited to transport charges effectively through the bridge to the

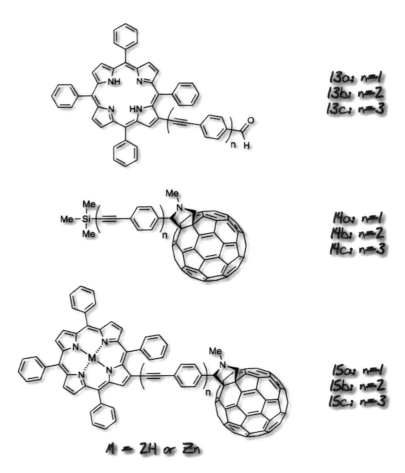

Fig. 9.16 Schematic representation of the H$_2$P/ZnP–oPPE$_n$–C$_{60}$ triads and their appropriate reference compounds

electron-accepting fullerene. This led us to validate the conformation of the bridge, namely the relative positions of the phenyl rings. A planar configuration, for instance, favors the preservation of the π-conjugated path. On the other hand, orthogonally arranged phenyl rings inevitably interrupt the extension of the π-conjugated system. Simply, an orbital overlap between the phenyl subunits of the molecular bridges should be envisaged. Matching of the orbital-energy levels, namely matching the orbital energies (or alternatively the ionization potentials and electron affinities) of the single components (i.e. donor, bridge and acceptor) is pivotal. In other words, for efficient electron transfer through molecular bridges, the energy levels of donors, bridges and acceptors must match each other in order to allow the electrons to be transferred from one building block to the other without any energy-penalty. No doubt, energies determine indisputably the

9.1 p-Phenyleneethynylene Molecular Wires

mechanism of the charge-transfer process. Therefore, the free-base porphyrin and the corresponding zinc complex were studied in light of the influence of structural properties on their electron-transfer features.

9.1.2.1 Absorption Studies

Similarly to the previously described systems, namely $exTTF-oPPE_n-C_{60}$, the absorption characteristics of **15a–c** and their reference compounds **13a–c** and **14a–c**, point to a lack of electronic communication between the building blocks in the ground state. The spectral signatures of C_{60}, the oPPE units and the porphyrins are preserved in **15a–c**. As an example, let us inspect the ground-state absorption of **13c** and **15c** (Fig. 9.17). The spectra are dominated by the porphyrin absorption with Soret bands around 430 nm and Q-bands at 560 and 590 nm. The oPPE absorption maximizes around 350 nm. Interestingly, a comparison between the oPPE absorption in **13c** and **14c** reveals a blue-shift of the latter absorption, which is due to the functionalization. Replacement of the terminal aldehyde groups in **13** by an sp^3 carbon from the fulleromethylpyrolidine results in a loss of conjugation length. This displays the sensibility of the electronic properties on small structural modifications. Summarizing, no significant alterations (i.e. shifts or changes) of the characteristic spectral features of the individual building blocks were observed in the conjugates.

These findings have been further corroborated by cyclic voltammetry studies. Similar results when compared with those of the electrochemical experiments with $exTTF-oPPE_n-C_{60}$ were gathered. Particularly, the reduction and oxidation

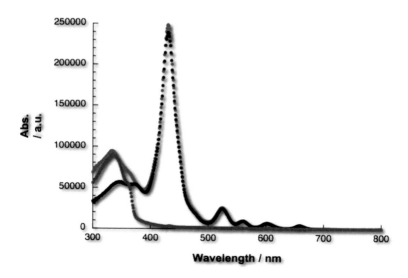

Fig. 9.17 UV–visible absorption spectra of **13c** (*black*), **14c** (*grey*) and **H₂P – 15c** (*blue*) in toluene solution

potentials of the building blocks **13a–c** and **14a–c** bear close resemblance with H_2P/ZnP–oPPE$_n$–C_{60} (**15a–c**). At this point, we assume that the energy levels, i.e. the orbitals, of the building blocks match favorably each other and provide suitable conditions for efficient transfer of charges in H_2P/ZnP–oPPE$_n$–C_{60}. In order to ratify this hypothesis, it was necessary to inquire into the influence of structural and electronic properties of these systems. This has been accomplished by means of quantum mechanical methods.

9.1.2.2 Molecular Modeling

Quantum chemical studies on the previously mentioned oPPE molecular wire systems led to the conclusion that rotations around the phenyl rings of these π-conjugated oligomers are not restricted due to very low rotational barriers (<2.0 kcal/mol). However, the dihedrals between the phenyl rings tend to adopt nearly planar configurations, which implies extended π-conjugation throughout the entire oPPE linker. Taking these findings into account, we have carefully looked at the rotational barrier between the phenyl rings of the bridge in **15a–c** (Fig. 9.18). For that reason, the geometries of the zinc porphyrin (ZnP) and the free-base porphyrin (H_2P) derivatives were optimized at different levels of theory, namely semi-empirically

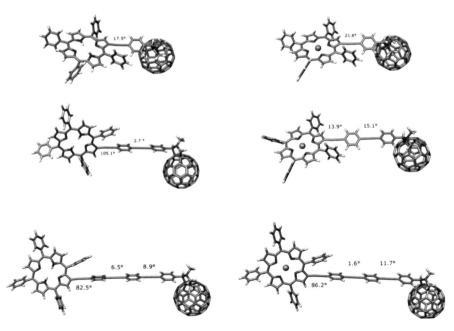

Fig. 9.18 Optimized geometries of the investigated triads **15a–c** including the dihedral angles between the phenyl rings and the bridge and the porphyrin moiety. *Left column*: free-base porphyrin derivatives. *Right column*: zinc porphyrin derivatives

($AM1^*$ and $PM3$ [12]) and by density functional theory (DFT) methods. To evaluate the rotational barriers, further single-point DFT calculations with improved basis sets were employed on structures with systematically varying dihedral angles and compared to the optimized geometries. Interestingly, both monomeric derivatives **15a**, i.e. ZnP and H_2P, show a dihedral angle of approximately 20°, leading to close contacts between the porphyrins and the C_{60} cage (Fig. 9.18). This might affect the excited-state deactivation rate of the monomer **15a** in comparison to the dimer **15b** and trimer **15c** by making a rapid through-space electron-transfer reaction feasible. In contrast to **15a**, the two remaining oligomers contain nearly planar bridge configurations with dihedral angles in the range of ± 14°, which is in good agreement with our previous results (see Sect. 9.1.1). Enthrallingly, in the trimer **15c** the orientation of the adjacent phenyl ring and the porphyrin framework turned out to be roughly orthogonal, which might have a strong impact on the conjugation between the bridge and the porphyrins (Fig. 9.18). Replacing the free-base by the metallated porphyrin enhances this effect on account of increasing electron density stemming from the metal ion. Torsions around the ethynylene bond attached to the porphyrin apparently become more and more restricted with increasing length of the oPPE linker, which may impact the electron-transfer properties of these triads. Thus, a lack of coplanarity between the porphyrin and oPPE, as is evident in the trimer **15c**, might per se lead to a loss of electronic communication between donor and acceptor and thus to slower charge-recombination rates.

In order to complement these findings, frontier orbitals schemes were analyzed. In particular, the HOMO/LUMO orbitals clearly reflect the donor–acceptor character of the systems. In line with our expectations, the HOMO and LUMO orbitals are strongly localized on the donor and the acceptor, i.e. the porphyrins and the fullerene, respectively (Fig. 9.19). Significant overlap between the donor and the bridge suggests strong electronic coupling between donor and acceptor and facilitates electron injection from the excited porphyrin to the oPPE connector. However, in the ZnP trimer, coefficients of the HOMO are significantly present in the bridge structure and, hence, imply a stronger donor character of the ZnP derivatives in comparison to H_2P. Considering charge-separation processes in these molecules, a stronger donor character will unambiguously lead to higher charge-separation rates. A possible explanation for these findings is that the linker of the ZnP derivative needs to accommodate higher electron density from the central metal ion. For this reason, electron density is redistributed into the adjacent vacant levels of the bridge. As a consequence, Coulomb repulsive interactions force the zinc porphyrin into an orthogonal arrangement to the phenyl units of the bridge. In the monomer, on the other hand, the short separation distance allows a facile transfer of electron density onto the LUMOs of the fullerene core.

On the other hand, the electron-accepting features of the fullerene, as incorporated into the triads, are perfectly preserved. All oligomers exhibit a localization of the LUMO exclusively surrounding the symmetrical C_{60} cage. Furthermore, the short distance between donor and acceptor in the monomer **15a** leads to significant overlap between HOMO and LUMO, giving rise to extremely fast electron-transfer processes.

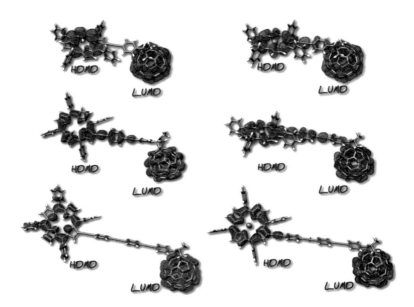

Fig. 9.19 HOMO/LUMO orbital schemes as resulted from DFT optimizations displaying the donor–acceptor character of the triads. The HOMOs are represented in *orange–red* and LUMOs in *green–blue*

Taking the results from the absorption and electrochemical studies into account (see above), it was necessary to evaluate the relationship between the energy levels of the pristine building blocks. For this reason, we have optimized (DFT) the structures of the reference compounds, i.e. **13a–c** and **14a–c**, and their pristine building blocks methylfulleropyrrolidine **1**, ZnP, H$_2$P, and *o*PPEs in order to investigate their HOMO/LUMO orbital energies. Thereby, important evidence for the interactions between the donor, bridge and acceptor sites was found.

Figure 9.20 relates the HOMOs and LUMOs of the building blocks and reference compounds to the corresponding orbitals in the triads. Herein, the significance of the matching of the orbital energy levels becomes apparent. The most interesting subject of concern are the LUMO energies in the triads **15a–c** and C$_{60}$–*o*PPE references **14a–c**, namely −3.28 eV, which exactly matches the LUMO energy of the methylfulleropyrrolidine reference **1** and clearly demonstrates the role of the C$_{60}$ core as the predominant electron acceptor. Importantly, this feature is evident in all triads independent of the distance between donor and acceptor. On the other hand, a comparison of the HOMO energies of the triads and the porphyrin–*o*PPE conjugates **13a–c** to those of pristine H$_2$P and ZnP reveals the electron-donating features of the porphyrins. Here, the porphyrin HOMOs can be energetically compared to the corresponding HOMOs in **13a–c** and **15a–c**. Thus, attaching the *o*PPE linker to the porphyrin derivatives does not alter the HOMO energy at all. Equally, the HOMO/LUMO energies of the donor and acceptor are preserved in all triads, leading to the assumption that the electronic communication between the two does

9.1 p-Phenyleneethynylene Molecular Wires

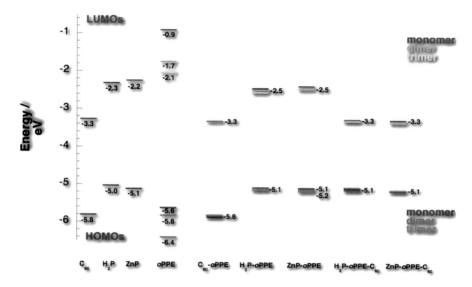

Fig. 9.20 HOMO/LUMO orbital energies as resulted from DFT optimizations of the building blocks of the triads in relation to the triads themselves. The HOMO energies are represented in *red* and LUMO energies in *green*

not depend on the bridge length. Apparently, the bridge orbitals do not mix with those of the donor and the acceptor, which implies perfect conditions for a superexchange rather than a through-bond hopping mechanism for the electron transfer processes. Nevertheless, the presence of significant overlap between the oPPE linkers and the electron donor and acceptor render the possibility of a thermally induced hopping mechanism involving the bridge orbitals at higher temperatures. Firstly, the energy difference of the orbitals of the oPPE oligomers vanishes upon covalent binding to the porphyrins. Implicit extension of the π-system of the porphyrins onto the bridge is a possible explanation. On the other hand, in the case of the fullerene, the HOMO and LUMO orbitals remain strongly localized on C_{60}, implying a lack of electronic perturbation by the bridge. Moreover, the HOMO–LUMO gap of the oPPE units is approximately 4.2 eV and hence more than 1.0 eV higher than in the corresponding porphyrin–oPPE conjugates **13a–c** and the triads **15a–c**. This implies strong electronic coupling between the porphyrins and C_{60} which does not depend on the length of the oPPE linker and leads to nearly distant-independent electron-transfer processes. Conclusively, the HOMO/LUMO energy schemes perfectly comply with the results obtained throughout the absorption studies, where it was shown that in the absorption spectra of the triads the spectral features of the building blocks are discernible.

Equally, the trends found in cyclic voltammetry measurements are perfectly reproduced by this computational study. For instance, the HOMO/LUMO energies are in perfect agreement with the oxidation and reduction potentials obtained from electrochemistry. It is noteworthy that electrochemical measurements suggest that

Fig. 9.21 Local electron affinity maps of the exTTF–oPPE$_3$–C$_{60}$ trimer compared to the H$_2$P/ZnP–oPPE$_3$–C$_{60}$, **15c**, triads, scaling high to low in *red* to *blue*

the reduction of the C$_{60}$ moiety in **14a–c** does not depend on the attached oPPE-bridge and thus rule out strong interactions between these two moieties. Our DFT calculations corroborate these features revealing that the reduction potentials of **14a–c**, which appear at around −3.8 eV (experimentally) comply with the LUMO energies of C$_{60}$ (calculated to −3.3 eV[2]). These values exceed the LUMO energies of oPPE by more than 1.0 eV. Along with the assumption that the calculated orbital energies match the potentials obtained by electrochemistry, these findings imply a lack of electronic interactions between the redoxactive moieties. For instance, the oxidation and reduction of **13a–c** are separated by an energy gap in the range of 2.2 and 2.3 eV, which agrees with the energy required for a HOMO → LUMO transition obtained from calculations, namely 2.6–2.7 eV (see Footnote). Similarly, the potential gaps for the redox processes in the triads **15a–c** that appear between 1.6 and 1.9 eV are ascribed to the calculated HOMO → LUMO transitions calculated to occur at 1.8 eV.

For the sake of comparison to the above mentioned exTTF–oPPE$_n$–C$_{60}$ systems, we have computed the local electron affinity of **15a–c**. Mapping the local electron affinity onto the electron density provides further insight into the postulated electron-transfer features of H$_2$P/ZnP–oPPE$_n$–C$_{60}$. The results for exTTF–oPPE$_n$–C$_{60}$ and **15c** are represented in Fig. 9.21. In our foregoing studies of exTTF–oPPE$_n$–C$_{60}$, the electron-transfer pathway through the bridge, i.e. the areas of high

---

[2] Calculated energies in vacuum are higher due to a lack of solvent stabilization.

local electron affinity, was disrupted by the triple bonds due to their strong polarizing features (see Sect. 9.1.1). In addition, electron density minima appeared between the phenyl rings. Currently, the electron density in the free-base and the zinc metallated derivative of **15c** is more homogeneously distributed throughout the whole bridge due to the more electron-rich porphyrin donors in comparison to exTTF. In line with the electron density, the local electron affinity exhibits a much more uniform distribution. Nevertheless, minima are localized at the triple bond sites but the discrepancy between values of high (orange) and low (yellow to green) electron affinity is less pronounced due to higher overall electron density stemming from the electron-rich porphyrin donors (Fig. 9.21). Inevitably, these findings suggest improved electron-transfer features for the porphyrin donor derivatives in comparison to the exTTF donor derivatives of the DBA triads. Thus, the β values obtained should turn out to be lower than 0.20 Å^{-1}. Interestingly, the ZnP derivatives exhibit notably high electron affinity values at the central metal ion which, in turn, reflects the inhomogeneous distribution of the electrons around the Zn atom that facilitates electron injection into the bridge.

To summarize, quantum mechanical investigations of the above systems provide evidence for the occurrence of effective photoinduced charge-transfer reactions. Under these aspects, we will now move on to the photophysical characterization of the proposed processes and verify the charge-separation features by means of steady-state and time-resolved spectroscopic techniques.

9.1.2.3 Photophysics

Firstly, we draw our attention to the steady-state emission characteristics of the reference compounds. Since, the emission spectra of C_{60} and oPPE have already been discussed in the context of exTTF–oPPE$_n$–C_{60}, it is sufficient to mention the spectral range of the individual fluorescence. While the fullerene emission maximizes typically at 715 nm, the oPPE fluorescence is observed in the 400–500 nm range. The H_2P/ZnP features, on the other hand, are found between 600 and 700 nm with the highest quantum yields of all references (i.e. 0.2 vs. 6.0×10^{-4} for C_{60}). In a similar fashion, transient absorption spectra show distinct features for all compounds: Singlet-excited states are discernible between 500 and 700 nm for the oPPEs, at 480 nm for H_2P/ZnP and 880 nm for C_{60}. The metastable singlet-excited states are deactivated by fast intersystem crossing processes to the corresponding triplet states in all reference compounds. The triplet–triplet absorptions are all located in the range between 500 and 900 nm.

The photophysical behavior of **14a–c** matches the results presented in Sects. 8.1.1 and 9.1.1. Hence, it should only be mentioned that attaching C_{60} to the oPPEs results in a rapid transfer of singlet excited-state energy from the photoexcited oPPEs to the covalently attached C_{60}. At the conclusion of this energy-transfer scenario stands the successful formation of the C_{60} triplet state. It infers that the underlying reaction sequence involves besides the energy transfer an intersystem crossing from the C_{60} singlet to the energetically lower-lying triplet manifold.

Now, when turning our attention to H$_2$P/ZnP–oPPE$_n$–C$_{60}$ (**15a–c**) we should keep in mind that the aforementioned energy transfer from the oPPEs to C$_{60}$ in the reference compounds **14a–c** is nearly quantitative. Thus we focus on the fate of the singlet excited state at the fullerene end. Steady-state fluorescence experiments suggest that the general reaction scheme is not affected in the presence of H$_2$P or ZnP (i.e., H$_2$P/ZnP–oPPE$_n$-C$_{60}$). In case of selective excitation of the oPPE part in the triads, which is achieved by $\lambda_{exc} = 300$–500 nm, the rapid intramolecular transduction of excited-state energy from the oPPEs to C$_{60}$ is still perceptible. The outcome of this energy transfer process is the corresponding singlet excited state of C$_{60}$ generated with nearly unity quantum yields as evidenced by fluorescence quantum yields of $\sim 5.0 \times 10^{-4}$ which are comparable to the quantum yield of pristine C$_{60}$ (i.e. 6.0×10^{-4})—see Table 9.3.

The presence of H$_2$P and ZnP in the triads **15a–c** is reflected by their strong emission, which dominates large parts of the steady-state fluorescence spectra. However, in comparison to the H$_2$P/ZnP references the porphyrin centered fluorescence is appreciably quenched in all H$_2$P/ZnP–oPPE$_n$–C$_{60}$. Interestingly, this

Table 9.3 Selected spectroscopic data for the investigated triads **15a–c** and the appropriate reference compounds

	ϕ_{fl}	$\tau_1 [ns]$	$k_{CS} [s^{-1}]$	$k_{CR} [s^{-1}]$
C$_{60}$-reference	6.0×10^{-4}	1.2	–	–
H$_2$P-reference	1.1×10^{-1}	9.8	–	–
ZnP-reference	4.0×10^{-2}	2.4	–	–
15a (H$_2$P/ZnP)	Toluene: 1.0×10^{-2}/ 7.8×10^{-4} THF: 5.8×10^{-3}/ 3.9×10^{-4} Benzonitrile: 4.6×10^{-3}/ 2.3×10^{-4}	<0.1/<0.1	8.50×10^{8}/ 2.50×10^{9}	4.88×10^{6}/ 4.90×10^{6}
15b (H$_2$P/ZnP)	Toluene: 6.1×10^{-2}/ 7.4×10^{-3} THF: 4.4×10^{-2}/ 2.7×10^{-3} Benzonitrile: 3.2×10^{-2}/ 2.1×10^{-3}	4.18/0.45	2.39×10^{8}/ 2.22×10^{9}	1.15×10^{6}/ 1.44×10^{6}
15c (H$_2$P/ZnP)	Toluene: 9.1×10^{-2}/ 2.7×10^{-2} THF: 8.1×10^{-2}/ 1.7×10^{-2} Benzonitrile: 7.9×10^{-2}/ 1.3×10^{-2}	5.97/1.18	1.68×10^{8}/ 0.85×10^{9}	0.84×10^{6}/ 1.10×10^{6}
14a	$\sim 5.0 \times 10^{-4}$	1.2 ± 0.02	–	–
14b	$\sim 5.0 \times 10^{-4}$	1.4 ± 0.02	–	–
14c	$\sim 5.0 \times 10^{-4}$	1.4 ± 0.02	–	–

9.1 p-Phenyleneethynylene Molecular Wires

occurs regardless of the excitation wavelength. In other words, the quenching is independent on the excitation (i.e. Soret or Q bands), which infers that the presence of C_{60} in the triads must enhance the fluorescence deactivation. The quenching of the porphyrin fluorescence in THF is represented in Fig. 9.22.

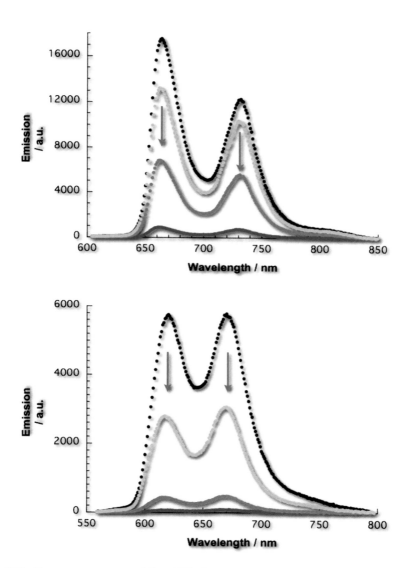

Fig. 9.22 Fluorescence spectra of C_{60}–oPPE–H_2P and H_2P—*upper part*—and C_{60}–oPPE-ZnP and ZnP—*lower part*—in THF upon 460 and 440 nm excitation, respectively, with solutions displaying the same optical density at the excitation wavelengths (i.e. 0.1). H_2P and ZnP spectra are represented in *black* and the triads **15a** in *blue*, **15b** in *light blue*, and **15c** in *violet*

To gain further insight into the nature of the intramolecular deactivation processes and their products, complementary time-resolved fluorescence and transient absorption measurements deemed necessary. From time-resolved emission studies, the fluorescence lifetimes were obtained (Table 9.3). Inspecting the lifetime of the H_2P/ZnP emission in the triads reveals a notable shortening in comparison with the porphyrin references. Further, upon varying the solvent polarity from toluene ($\epsilon = 2.4$) to benzonitrile ($\epsilon = 24.8$) the lifetimes decrease. This trend is in line with the gradual intensification of the fluorescence quenching upon increasing solvent polarity as evidenced by decreasing quantum yields in Table 9.3. Such a solvent dependence infers that the deactivation occurs via an intramolecular electron transfer between the porphyrin donors and the C_{60} acceptor. Accordingly, the generated radical ion pairs, namely $H_2P^{\bullet+}/ZnP^{\bullet+}-oPPE_n-C_{60}^{\bullet-}$, are better stabilized by a solvent of higher polarity, which is reflected by shorter fluorescence lifetimes.

Transient absorption measurements focused on the generation of the singlet excited states of H_2P and ZnP because the overlapping absorptions of all building blocks in the conjugates would cause any other analysis to become equivocal. Nevertheless, in compliance with previous work on $exTTF-oPPE_n-C_{60}$ (see Sect. 9.1.1) we must assume that exciting C_{60} will lead to equal reaction patterns. Rather than observing the slow intersystem crossing dynamics present in the H_2P/ZnP references, the deactivation of the singlet–singlet absorption of the porphyrins in **15a–c** is accelerated. Importantly, the singlet–singlet decay rates quantitatively match the values derived from the fluorescence experiments. However, at the conclusion of the decay the transient absorption spectra are lacking any triplet excited-state signatures. To the same extent, varying the solvent polarity from toluene to benzonitrile leads to an increase of the singlet deactivation rate. Thus, instead of the aforementioned triplet excited state of C_{60}, the product of the singlet–singlet deactivation in the conjugates is presumed to be of charge-transfer character. Differently speaking, photoexcitation of **15a–c** leads to the formation of radical ion pairs, $H_2P^{\bullet+}/ZnP^{\bullet+}-oPPE_n-C_{60}^{\bullet-}$.

In fact, it was possible to prove the formation of the radical ion pair state by transient absorption spectroscopy. Particularly, at the expense of the vanishing H_2P/ZnP singlet absorption new features with maxima in the 600–700 nm range as well as at 480 nm grow in. These maxima correspond to the one-electron oxidized π-radical cations of H_2P ($H_2P^{\bullet+}$) and ZnP ($ZnP^{\bullet+}$). Additionally, in the near-infrared region the spectral signatures of the one-electron reduced anion of C_{60} are discernible at 1000 nm (Fig. 9.23).

Interestingly, the lack of charge transfer in $C_{60}-oPPE$ suggests that, the transient absorption data of **15b** and **15c** should be interpreted as charge separation occurring in a single step. Therefore, the involvement of an intermediate step, in which the oligomeric $oPPE$ bridge would be oxidized ($H_2P/ZnP-oPPE_n^{\bullet+}-C_{60}^{\bullet-}$) is ruled out. This, in turn, corroborates the results from the quantum mechanical investigation, which proposed a coherent superexchange mechanism due to the structural properties in the conjugates at short donor–acceptor separation distances.

The charge-transfer dynamics are evaluated on the basis of the growth and decay kinetics of the C_{60} π-radical anion and of the H_2P/ZnP π-radical cations

9.1 p-Phenyleneethynylene Molecular Wires

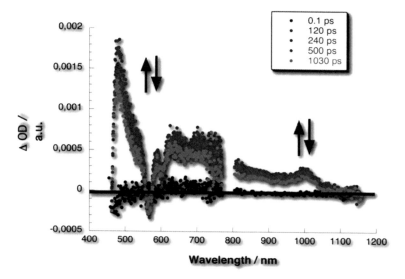

Fig. 9.23 Differential absorption spectra (visible and near-infrared) obtained upon femtosecond flash photolysis (387 nm–200 nJ) of **Zn15b** in argon saturated THF with several time delays between 0 and 1000 ps at room temperature—illustrating the charge transfer

signatures as shown in Fig. 9.24. On the picosecond time-scale, both spectral attributes are stable and only start to decay slowly when moving to the nanosecond time regime. Inspecting the time absorption profiles confirms the single-step decay dynamics of the radical ion pair states in **15a–c**. The charge-recombination rates were easily determined from complementary nanosecond experiments. In particular, we fitted the decays of the $H_2P^{•+}/ZnP^{•+}$ and $C_{60}^{•-}$ fingerprints to mono-exponential rate laws.

Taking the aforementioned solvent dependence into account, namely the stabilizing effects of polar solvents on the radical ion pairs, we conclude that the charge-recombination dynamics are located in the inverted region of the Marcus parabola. For that reason, we have calculated the driving forces for charge separation and charge recombination in each of the three solvents according to the method by Weller [13].

Table 9.4 represents the calculated $-\Delta G$ values for the charge separation and charge recombination processes. Hereby, the charge recombination falls into the inverted regime of the Marcus parabola. With these values in hand, it was possible to place the different possible reaction pathways in a state diagram (Fig. 9.25).

Thus, from fluorescence lifetime and transient absorption measurements we gathered the electron-transfer rate constants, i.e. both for charge-separation and for charge-recombination. Next, we plotted these rate constants as a function of donor–acceptor distance. From the resulting linear dependence (Fig. 9.26) it is possible to determine the attenuation factors β for the presented donor–acceptor ensembles as 0.11 Å^{-1}.

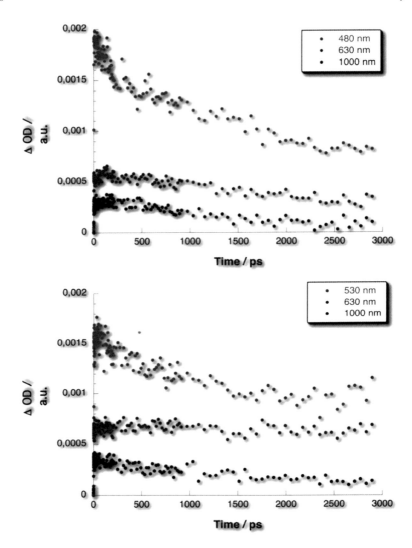

Fig. 9.24 *Upper part*—time–absorption profiles at 480, 640 and 1000 nm of the differential absorption measurements with **Zn15b** in THF solutions—*lower part*—time–absorption profiles at 530, 630 and 1000 nm of the differential absorption measurements with **Zn15c** in THF solutions—monitoring the charge transfer processes

9.1.2.4 Summary

In view of the photoinduced processes seen in exTTF–oPPE$_n$–C$_{60}$ (see Sect. 9.1.1), the replacement of exTTF in the presented H$_2$P/ZnP–oPPE$_n$–C$_{60}$ (**15a–c**), leads only to minor alteration in the overall reaction pattern. Upon photoexcitation the systems undergo intramolecular electron transfer from the electron donating ZnP/H$_2$P to the

9.1 p-Phenyleneethynylene Molecular Wires

Table 9.4 Solvent dependent driving forces for charge separation (CS) out of the porphyrin singlet excited state and charge recombination (CR) to the ground state/porphyrin triplet excited state calculated after the dielectric continuum model (dielectric constant ϵ: toluene 2.4; THF 7.6; oDCB 9.8, benzonitrile 24.9). The case, where charge recombination to the porphyrin triplet state is prohibited, is assigned as "n.p."

	$E_{0\rightarrow 0}$ [eV]	E_T [eV]	Solvent	$-\Delta G_{CS}$ [eV]	$-\Delta G_{CR}$ [eV]	
					Ground state	Triplet state
15a (H$_2$P/ ZnP)	1.89/2.04	1.40/1.53	Toluene:	−0.03/0.22	1.91/1.82	0.51/0.29
			THF:	0.32/0.57	1.57/1.47	0.17/n.p.
			Benzonitrile:	0.43/0.68	1.46/1.36	0.06/n.p.
15b (H$_2$P/ ZnP)	1.89/2.04	1.40/1.53	Toluene:	−0.31/−0.06	2.20/2.10	0.80/0.57
			THF:	0.23/0.48	1.66/1.56	0.26/0.03
			Benzonitrile:	0.40/0.65	1.49/1.39	0.09/n.p.
15c (H$_2$P/ ZnP)	1.89/2.04	1.40/1.53	Toluene:	−0.58/−0.18	2.47/2.22	1.07/0.69
			THF:	0.04/0.44	1.85/1.60	0.45 /0.07
			Benzonitrile:	0.24/0.64	1.65/1.40	0.25/n.p.

electron-accepting C$_{60}$. The lifetimes of the charge-separated states are in the order of hundreds of nanoseconds and tend to increase with increasing distance between the two redoxactive units. In **15c**, charge separation occurs over a distance of up to 23 Å, implying a through bond mechanism, where the bridge plays an important role. On the other hand, in **15a** and **15b** electron transfer via a superexchange mechanism seems to be the most probable operative mode. Quantum chemical investigations provided insight into the electronic and structural properties of the investigated systems. Analysis of the electronic structure revealed a strong localization of the HOMOs and LUMOs on the electron-donating ZnP/H$_2$P and electron-accepting C$_{60}$, respectively. In comparison with the *ex*TTF derivatives, ZnP/H$_2$P seem to exhibit better electron-donating character, which was also reflected by lower attenuation factors with values of 0.11 Å^{-1} for H$_2$P/ZnP–*o*PPE$_n$–C$_{60}$ versus 0.20 Å^{-1} for *ex*TTF–*o*PPE$_n$–C$_{60}$. Supplementary evidence for the improved electron-transfer features came from electron affinity calculations. These reveal indeed a more homogeneous distribution of electron density and local electron affinity. Furthermore, strong electronic coupling between ZnP/H$_2$P and C$_{60}$, which is invariant from the length of *o*PPE leads to nearly distance-independent electron-transfer processes. Extremely fast charge-separation dynamics in **15a** and fast charge-recombination dynamics in **15c**, on the other hand, are due to geometrical peculiarities of the systems. A schematic representation of the possible reaction pathways is given in the energy diagram of Fig. 9.25.

9.1.3 Meta-Connectivity—Influence of Structure on Molecular Wire Properties

At the conclusion of the investigation of molecular wire properties of *oligo-(para-phenyleneethynylene)*s (*o*PPEs), the properties of structurally modified

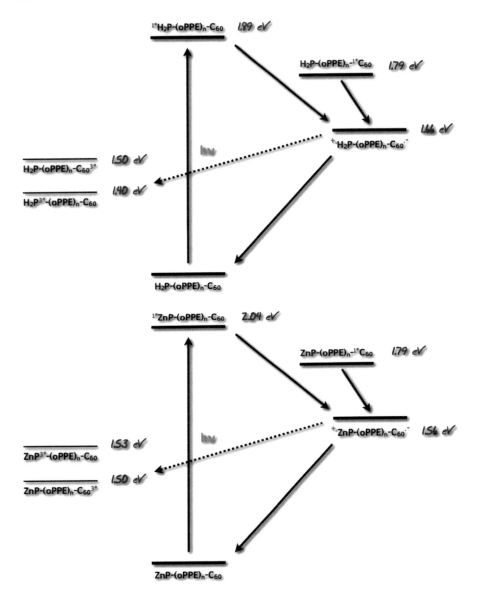

Fig. 9.25 Schematic illustration of the reaction pathways in photoexcited **H₂P15b** (*top*) and **ZnP15b** (*bottom*) with the state energies as determined in THF. The different deactivation pathways are indicated by *arrows*

9.1 p-Phenyleneethynylene Molecular Wires

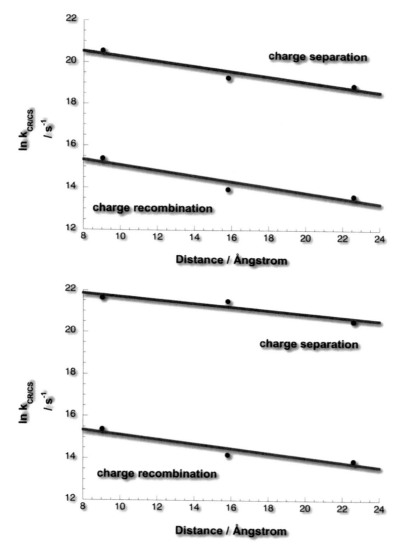

Fig. 9.26 Edge-to-edge distances (R_{EE}) dependence of charge-separation (ln k_{CS}) and charge-recombination (ln k_{CR}) rate constants of H_2P–oPPE$_n$–C_{60}—upper part—and ZnP–oPPE$_n$–C_{60}—lower part—triads ($n = 1, 2, 3$) in nitrogen saturated THF at room temperature. The slope represents β with the value of 0.11 Å^{-1}

oligo-phenyleneethynylene wires should be mentioned, namely the *meta*-connected isomers *oligo*-(*meta*-phenyleneethynylene)s (*o*MPEs). In this context, a series of $exTTF$–oMPE$_n$–C_{60} systems with up to four oligomeric bridge units and the appropriate reference compounds have been synthesized. The structures are represented in Fig. 9.27.

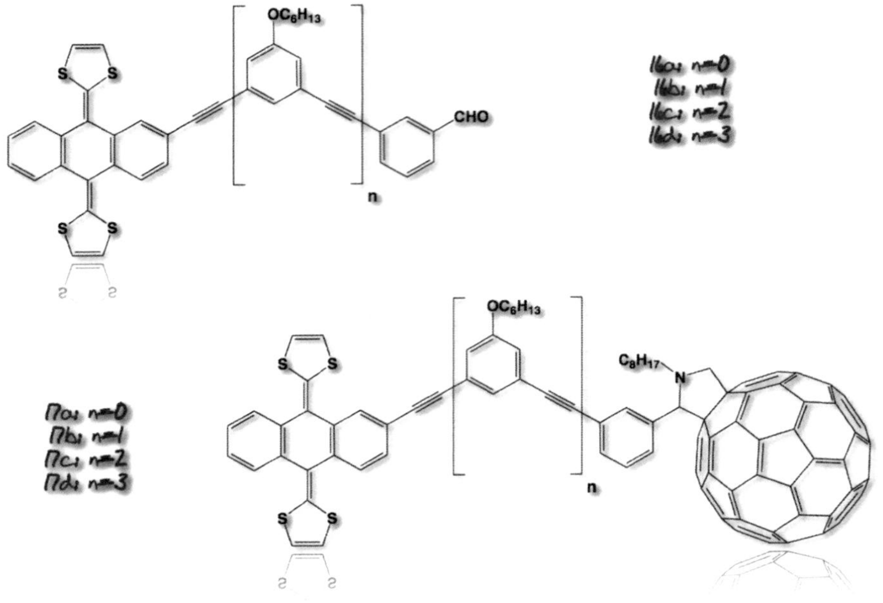

Fig. 9.27 Schematic representation of the $exTTF$–$oMPE_n$–C_{60} donor–acceptor triads and their appropriate reference compounds

The systems have been investigated in view of their photophysical characteristics and furthermore in view of aggregation phenomena, which turned out to be appealing throughout the photphysical studies. Any differences in electronic properties as they may be manifested between the *para-* and *meta-*connected systems are mainly governed by the more compact structure of the latter. The resulting ground-state features of both isomers, on the other hand, are comparable. Thus, the main focus of this subchapter is directed to the examination of aggregation phenomena.

9.1.3.1 Photophysics

Comparing the ground-state absorption features of oPPE and oMPE triads (Fig. 9.28), the main difference can be found in the absorption of the linkers, i.e. in the 350–550 nm region. In the *meta-*isomers the absorptions exhibit a distinctive pattern with maxima that hardly shift to the red part of the spectrum. This implies weaker or even a lack of π-conjugation relative to the corresponding *para* systems. For that reason, C_{60} (380 and 434 nm) and $exTTF$ (450 nm) absorption features are clearly distinguishable.

Turning to steady-state fluorescence studies, appreciable quenching of C_{60} emission at 710 nm was only discernable for the monomer **17a**, i.e.

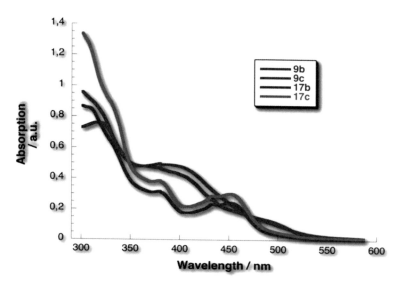

Fig. 9.28 UV–visible absorption spectra of **9b,c** (*blue/violet*) and **17b,c** (*red/dark red*) in *o*DCB solution

$exTTF$–$oMPE_1$–C_{60}. The quenching in the dimer **17b**, trimer **17c**, and tetramer **17d** is notably weaker than in **17a**. Moreover, it is invariant with respect to the $oMPE$ length (Fig. 9.29a). For the monomer we have calculated a quantum yield of 0.7×10^{-4}, whereas values of $\leq 3.0 \times 10^{-4}$ were derived for the remaining oligomers. Inspecting the quenching of the oligomer emission in **17a–d** with respect to **16a–d** resembles the situation found for the C_{60} emission (Fig. 9.29b). Such behavior prompts to an intramolecular electron-transfer deactivation process. In contrast to the electron-transfer processes in $exTTF$–$oPPE_n$–C_{60}, any charge-transfer in $exTTF$–$oMPE_n$–C_{60} seems to be governed by through-space interactions between the electron-donating $exTTF$ and the electron-accepting C_{60}. The accelerated formation of the radical ion pair state in **17a** is most likely due to very short distances between $exTTF$ and C_{60} resulting from the *meta* connectivity pattern.

Transient absorption measurements corroborated the generation of radical ion pairs in all oligomers **17a–c**. In particular, we have chosen 387 nm as excitation wavelength to photoexcite the ground state of $exTTF$ and C_{60}. Analogously to what has been seen in the corresponding *para*-substituted systems, the instantaneous grow-in of the C_{60} singlet excited-state absorption at 880 nm attests to the successful fullerene excitation. Moreover, instead of a slowly intersystem crossing, which afforded the corresponding triplet manifold of C_{60}, the singlet–singlet absorptions decay in the presence of $exTTF$ with very fast dynamics, i.e. 30 ps. Subsequently, new transients bearing strong maxima in the visible range at 680 nm and in the near-infrared at 1000 nm evolve (Fig. 9.30). As we already know, these spectral signatures correspond to the one-electron oxidized $exTTF$ radical cation

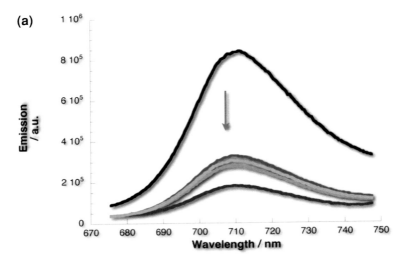

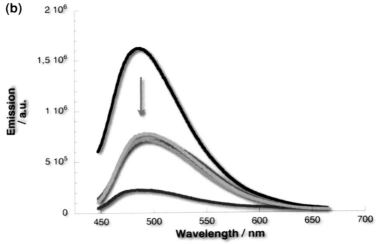

Fig. 9.29 a Fullerene emission of C_{60} reference **1** (*black*), **17a** (*blue*), **17b** (*light blue*), **17c** (*turquoise*) and **17d** (*light turquoise*) in *o*DCB, with matching absorption of 0.2 at the 375 nm excitation wavelength, **b** oligomer emission of **16c** (*black*), **17a** (*blue*), **17b** (*light blue*), **17c** (*turquoise*) and **17d** (*light turquoise*) in *o*DCB, with matching absorption of 0.2 at the 375 nm excitation wavelength

and the one-electron reduced C_{60} radical anion, respectively. In summary formation of an intramolecular radical ion pair state is confirmed.

Kinetic analyses of the formation of the radical ion pair state—formed through bond in **17a** and through space in **17b**, **17c**, and **17d**—revealed that the latter are meta-stable on the femto-/picosecond time scale. Hence, charge recombination

9.1 *p*-Phenyleneethynylene Molecular Wires

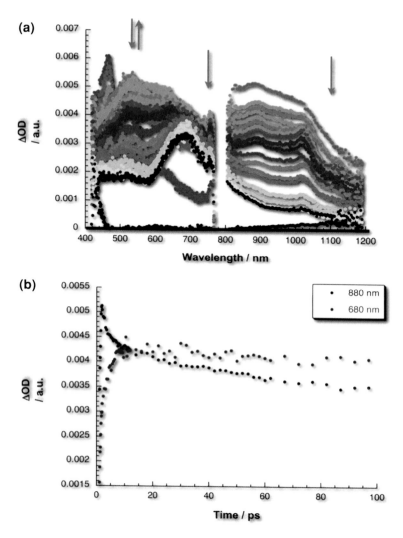

Fig. 9.30 a Differential absorption spectrum (visible and near-infrared) obtained upon femtosecond flash photolysis (387 nm) of 10^{-6} M^1 solutions of **17b** in nitrogen-saturated *o*DCB with time delays between 0 and 3000 ps at room temperature (*black* = 0 ps, *pink* = 1 ps, *yellow* = 2500 ps, and *black* = 2900 ps). **b** Time–absorption profiles of the spectra shown above at 680 and 880 nm to monitor the rapid decay of the singlet excited state of C_{60} and the simultaneous formation of the *ex*TTF radical cation

was examined on the nanosecond time scale upon excitation with a 6 ns laser pulse. The spectral signatures of the one-electron oxidized *ex*TTF radical cation and the one-electron reduced C_{60} radical anion—as detected immediately after the laser pulse—decay synchronously and give rise to kinetics that obey a clean unimolecular rate law. Lifetimes on the order of 30 ns were obtained for **17b**, **17c**, and **17d**, where through space interactions are assumed (Fig. 9.31), whereas through bond interactions in **17a** lead to lifetimes of 128 ns.

Conclusively, in the low-concentration regime (10^{-6} M^{-1}), where the electron donor-acceptor conjugates are present in their monomeric form, intramolecular electron-transfer processes, which are mainly governed by through-space interactions have been demonstrated by steady-state and time-resolved spectroscopic techniques.

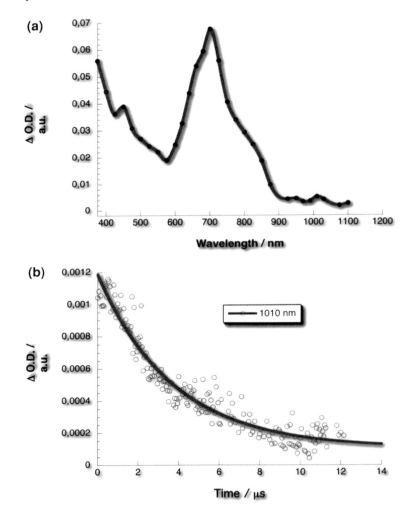

Fig. 9.31 **a** Differential absorption spectra (visible and near-infrared) obtained upon nanosecond flash photolysis (355 nm) of **17c** (2.0×10^{-6} M) in nitrogen-saturated oDCB solutions with a time delay of 100 ns at room temperature, indicating the radical ion pair state features at 680 and 1010 nm. **b** Time–absorption profiles of the spectra shown above at 1010 nm to monitor the decay of the radical ion pair state

9.1 p-Phenyleneethynylene Molecular Wires

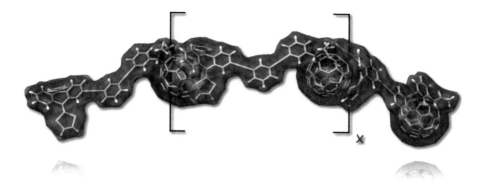

Fig. 9.32 Schematic representation of the aggregation due to π–π-stacking interactions in exTTF–oMPE$_n$–C$_{60}$ triads

By far more interesting is the aggregation of the oligomers, as discovered by atomic force microscopy (AFM) and absorption studies in the high concentration regime. In particular, AFM studies revealed aggregation of individual exTTF–oMPE$_n$–C$_{60}$ molecules, i.e. yielding (exTTF–oMPE$_n$–C$_{60}$)$_x$ aggregates (Fig. 9.32), due to π–π-stacking interactions between exTTF and C$_{60}$. These findings have been corroborated by mass spectroscopy. Hence, further insights into these organization phenomena regarding the electron-transfer processes were expected from spectroscopic studies.

To begin with, we should remind the reader that the lack of π-conjugation of the *meta*-substituted triads turned out to be an advantage in the subsequent spectroscopic assays. In other words, the fact that the only significant absorption stems from C$_{60}$ and exTTF opened the opportunity to test a large range of concentrations to study and follow the evolution of (exTTF–oMPE$_n$–C$_{60}$)$_x$ hybrids by spectroscopic means.

First insights into the complex formation came from absorption measurements. Herein, increasing the concentrations leads to a development of a distinct new absorption band centered red-shifted relative to the exTTF absorption at 470 nm. Notably, the C$_{60}$ and exTTF features still remain visible throughout the entire concentration range. Figure 9.33 represents the absorption of **17b** in toluene. Interestingly, in solvents with increasing polarity such as oDCB and benzonitrile the new developed features shift to 485 and 495 nm, respectively. This in turn, implies that the origin of the mentioned absorptions must be a charge-transfer character due to stabilizing effects in solvents of higher polarity.

For further understanding the nature of the new transition, we have turned to complementary steady-state emission experiments in the given concentration range. Several changes in comparison to the steady-state emission of the "non-complexed" systems are noted. For instance, with increasing concentrations at the expense of the broad and structureless exTTF emission a new band develops in the

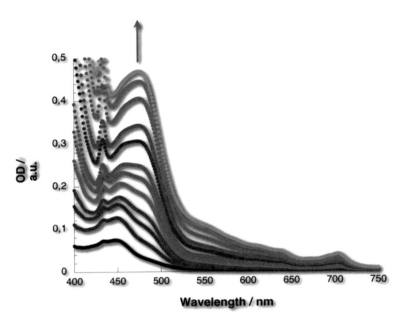

Fig. 9.33 Absorption spectra of a toluene solution of **17b** with increasing concentrations (5.5×10^{-6}, 9.3×10^{-6}, 1.2×10^{-5}, 1.3×10^{-5}, 1.6×10^{-5}, 1.8×10^{-5}, 2.0×10^{-5}, 2.4×10^{-5}, 2.8×10^{-5}, 3.4×10^{-5}, 3.9×10^{-5}, $4.3 \times 10^{-5} M$). *Arrows* indicate the progression of the titration

blue region—see Fig. 9.34. Considering the maximum of this band at 533 nm in toluene, the emerging emission is essentially a mirror image of the new features seen in the absorption measurements. Equally, a shift of the band can be observed upon variation of solvent polarity. In oDCB and benzonitrile the maxima are found at 550 and 575 nm, respectively. Thus, our steady-state absorption and emission studies insinuate that the features obtained upon increasing concentrations are due to intracomplex charge-transfer interactions.

To prove these findings and locate the origin of the new emission bands, we have employed complementary excitation experiments. The excitation spectra monitoring the 530 nm emission (Fig. 9.35) resemble the ground-state absorption characteristics of the conjugates. Conclusively, we presume that charge-transfer interactions in the sense of a partial redistribution of electron density from exTTF to C_{60} are responsible for the newly developing features.

Moreover, examining the quenching of the C_{60} emission in the concentration regime beyond 10^{-4} M causes a further decrease of fluorescence. This interesting trend implies that at higher concentrations an alternative deactivation pathway governs the photoreactivity of exTTF–oMPE$_n$–C_{60}. We assume tied charge-transfer interactions between C_{60} and exTTF due to the formation of intracomplex hybrids. Following the changes of both emission features, namely the development of the charge-transfer emission band and the decrease of C_{60} fluorescence as a

9.1 *p*-Phenyleneethynylene Molecular Wires

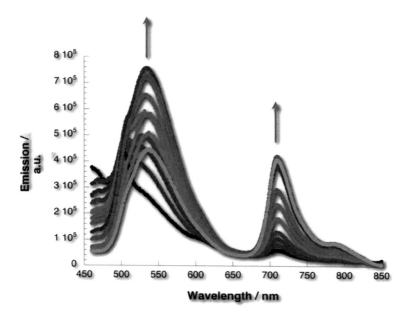

Fig. 9.34 Emission spectra of a toluene solution of **17b** with increasing concentrations (5.5×10^{-6}, 9.3×10^{-6}, 1.2×10^{-5}, 1.3×10^{-5}, 1.6×10^{-5}, 1.8×10^{-5}, 2.0×10^{-5}, 2.4×10^{-5}, 2.8×10^{-5}, 3.4×10^{-5}, 3.9×10^{-5}, $4.3 \times 10^{-5} M$). *Arrows* indicate the progression of the titration

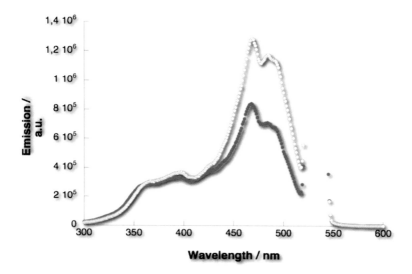

Fig. 9.35 Excitation spectrum of **17b** (*orange*) and **17c** (*pink*) in toluene solution monitoring the emission at 530 nm

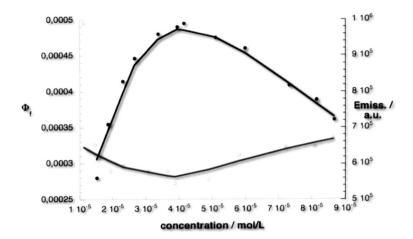

Fig. 9.36 Representation of the charge-transfer emission intensity (*black*) versus the C_{60} fluorescence quantum yield (*red*) in toluene solutions of **17b**

function of concentrations, reveal resembling trends. In other words, the charge-transfer band reaches its maximum at a concentration, when the C_{60} emission quenching is at its peak. This is illustrated in Fig. 9.36.

Investigating the transient absorption features at high concentrations leads to entirely different observations than those, which were detected at low concentrations. 482 nm excitation was chosen in order to excite directly into the charge-transfer features of the intracomplex hybrids. The formation of the radical ion pair state was evidenced immediately after laser excitation. The spectral fingerprints of the *ex*TTF radical cation at 680 nm and the C_{60} radical anion at 1000 nm evolve instantaneously (Fig. 9.37a). Interestingly, the formation of the intracomplex radical ion pairs occurs on a time scale of several picoseconds (Fig. 9.37b). On the other hand, the lifetime was determined to be remarkably long with a value of 2.1 ns. This confirms, that in contrast to the intramolecular radical ion pair state found at low concentration with stabilities in the picosecond range, the complex formation at higher concentrations leads to improved stability of the radical ion pair.

Additionally, in order to examine the charge-recombination dynamics we turned to complementary nanosecond transient absorption measurements. Once more, the spectral fingerprints of the radical ion pair state emerged immediately after the laser pulse and their decays yielded charge-recombination lifetimes in the order of 4.0 μs (Fig. 9.38).

9.1.3.2 Summary

At this point, it should be mentioned that testing the aggregation at the level of quantum mechanics is still in progress and will provide further insights into the electronic interactions between *ex*TTF and C_{60}. However, the presence of such was

9.1 p-Phenyleneethynylene Molecular Wires

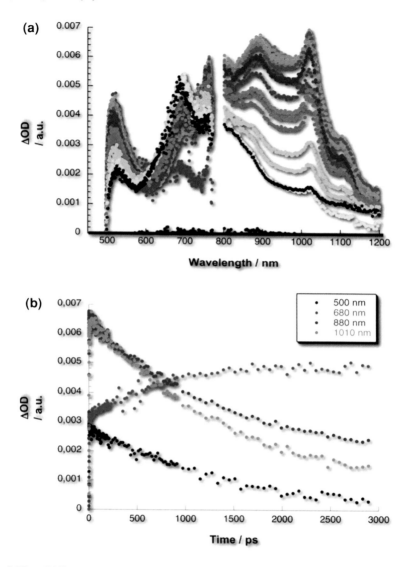

Fig. 9.37 a Differential absorption spectrum (visible and near-infrared) obtained upon femtosecond flash photolysis (482 nm) of 10^{-4} M^1 solutions of **17b** in nitrogen-saturated oDCB with time delays between 0 and 3000 ps at room temperature (*black* = 0 ps, *pink* = 1 ps, *yellow* = 2500 ps, and *black* = 2900 ps). **b** Time–absorption profiles of the spectra shown above at 500, 680, 880 and 1010 nm to monitor the formation of the radical ion pair state

already affirmed in complementary work dealing with *ex*TTF tweezers conjugates and pristine C_{60} molecules [14]. It has been demonstrated that the interactions between the aromatic system of *ex*TTF and C_{60} evoke the formation of a pincer-like 1:1 complex (Fig. 9.39). Furthermore, comparable intracomplex photoinduced

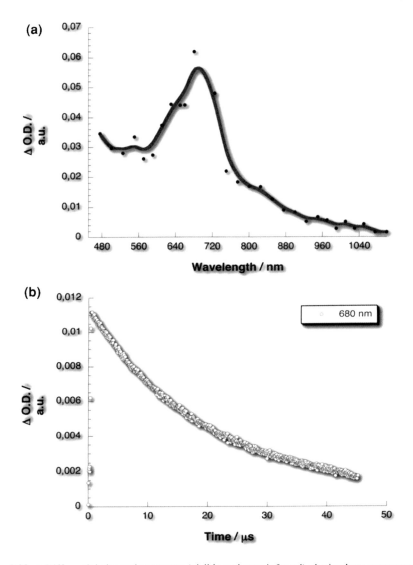

Fig. 9.38 **a** Differential absorption spectra (visible and near-infrared) obtained upon nanosecond flash photolysis (532 nm) of **17b** (1.2 × 10^{-4} M) in nitrogen-saturated oDCB solutions with a time delay of 200 ns at room temperature, indicating the radical ion pair state features. **b** Time–absorption profiles of the spectra shown above at 680 nm to monitor the decay of the radical ion pair state

charge-transfer processes between the electron donating exTTF moiety and the electron accepting fullerene occur in those donor–acceptor supramolecular π–π-complexes. The very short donor–acceptor distance consequences lifetimes of the charge-separated states that are very short (picosecond range). Under these aspects,

9.1 p-Phenyleneethynylene Molecular Wires

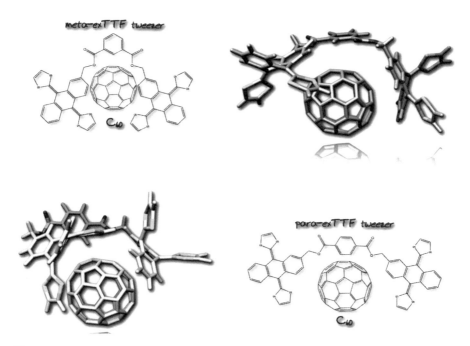

Fig. 9.39 Structural representation of the $exTTF:C_{60}$ 1:1 pincer-like complexes, demonstrating the effect of π–π-static interactions between the two aromatic systems

the herein presented complex formation between individual $exTTF$–$oMPE_n$–C_{60} molecules is rationalized by the presence of such π–π-interactions. Nevertheless, the long lifetimes of the $(exTTF$–$oMPE_n$–$C_{60})_x$ associates warrant further investigation. In particular, we need to determine whether the charge-transfer processes are truly intramolecular, i.e. involving the bridge, or intermolecular, i.e. between two $exTTF$–$oMPE_n$–C_{60} units. This work is still in progress.

For now, it has been shown, that the presence of attractive interactions as they are present between the two redox-active moieties, $exTTF$ and C_{60}, leads to aggregation phenomena in the high concentrations regime (i.e. $\geq 10^{-4}\,M^{-1}$). In turn, the photophysical response of the resulting intracomplex hybrids differs substantially from that found in the low concentrations regime, where only the monomeric form is present. In particular, the ground-state charge-transfer interactions result from a shift of electron density from $exTTF$ to C_{60} due to the short distance between donor and acceptor.

9.2 *oligo*-Fluorene Molecular Wires

In Sect. 8.1.2 we have already introduced *oligo*-fluorenes ($oFLs$) as molecular wires implemented into dumbbell architectures (i.e. C_{60}–oFL_n–C_{60}). It has been shown that *o*FL wires transduce energy upon photoexcitation when an acceptor, like a fullerene, is present. In view of the extensive photophysical study on *o*PPE molecular wires in the previous chapters, herein the results and differences in molecular-wire behavior between *o*FLs and *o*PPEs will be worked out. Since the photophysical investigation, in terms of steady-state absorption and emission studies as well as the time-resolved transient absorption and emission spectroscopic methods, does not differ from the previously presented experiments, we will focus on representative spectra.

9.2.1 ex*TTF*–o*FL*–C_{60} Donor–Acceptor Conjugates

In light of the excellent electron-donating properties of *ex*TTF and the favorable optical properties of *o*FL molecular wires, *ex*TTF–*o*FL$_n$–C_{60} ($n = 1, 2$) donor–acceptor conjugates have been synthesized as promising systems for photoinduced electron-transfer reactions (Fig. 9.40). The rather unusual features of fluorene-based oligomers, namely the persistence of the energy levels upon increasing number of oligomeric units, have already been outlined in Sect. 8.1.2. In that context, we should expect a rather weak distance dependence in the charge-transfer behavior of *ex*TTF–*o*FL$_n$–C_{60} DBA triads.

To probe the charge-transfer characteristics of these conjugates, we have carried out an extensive photophysical study and a number of quantum chemical calculation. Details on the synthesis and the electrochemical investigation are described elsewhere [15].

9.2.1.1 Photophysics

Let's first consider the ground-state features of **18a,b**. In line with the expectation, absorption studies in toluene, THF and benzonitrile confirm the lack of electronic communication between the building blocks, i.e. *ex*TTF, *o*FLs and C_{60}. Figure 9.41 shows the absorption spectra of **18a** and **18b** with their characteristic maxima at 430 nm (*ex*TTF), 345 nm (*o*FL) and 300 nm (C_{60}). Interestingly, a strong red-shift (20 nm) of the *o*FL absorption is envisioned when going from the monomer **18a** to the dimer **18b**. This is due to the extension of the π-conjugation length in bridge unit.

Further insight into the charge-transfer interactions came from complementary steady-state fluorescence studies. In line with the previously described C_{60}–*o*FL$_n$ and C_{60}–*o*FL$_n$–C_{60} conjugates, where the oligofluorene fluorescence was nearly quantitatively quenched—due to efficient transduction of singlet excited-state energies from the *o*FL units to C_{60}—similar trends evolve for *ex*TTF–*o*FL$_n$–C_{60}. Figure 9.42 represents the quantitative quenching of the *o*FL fluorescence in **18b**.

When compared to C_{60}–*o*FL$_n$ and C_{60}–*o*FL$_n$–C_{60}, drastic changes of the emission properties are observed upon inspecting the fullerene fluorescence

9.2 oligo-Fluorene Molecular Wires

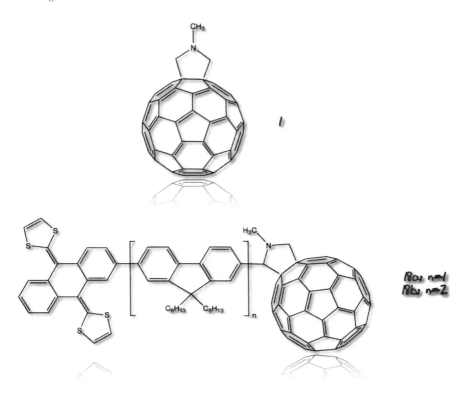

Fig. 9.40 Schematic representation of the $exTTF-oFL_n-C_{60}$ triads and the C_{60} reference compound

(Fig. 9.43). Hence, we can derive similar trends for the processes following photoexcitation in **18a,b** as in the corresponding $C_{60}-oPPE_n-exTTF$ triads. In particular, significant fluorescence quenching is evidenced for both oligomers, namely **18a** and **18b**. The quenching depends on the length of the oFL linker. With increasing length of the linker, the quenching of the C_{60} fluorescence is attenuated. The quantum yields document this trend with values of about 0.07×10^{-4} for the monomer and 0.90×10^{-4} for the dimer. In both cases the quenching is more than one order of magnitude relative to the quantum yields for the C_{60} references (i.e. 6.0×10^{-4}).

Next, a strong solvent dependence emerges. When varying the solvent polarity from, for example, toluene to benzonitrile, the C_{60} emission quenching increases (Fig. 9.44). In summary, such observations imply charge-transfer interactions between the C_{60} electron acceptor and $exTTF$ electron donor. Obviously, the charge transfer pathway passes the transiently formed C_{60} singlet excited state.

To prove these hypotheses, time-resolved measurements were employed. Transient absorption measurements on the femto- and nanosecond time-scale, i.e.

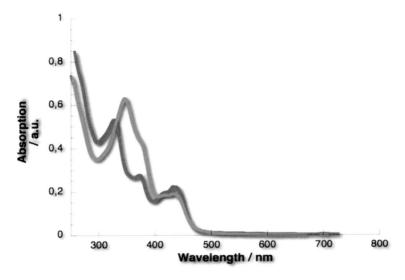

Fig. 9.41 Absorption spectra of **18a** (*red*) and **18b** (*orange*) displaying the maxima of the separate building blocks, i.e. 300 nm (C_{60}), 345 nm (*o*FL) and 430 nm (*ex*TTF)

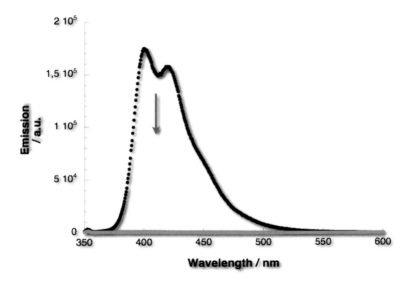

Fig. 9.42 Room temperature fluorescence spectra of a oFL$_2$ reference compound (*black*) and C_{60}–oFL$_2$–exTTF **18b** (*orange*) in THF displaying the quantitative quenching of the oligofluorene emission in the DBA triads

9.2 oligo-Fluorene Molecular Wires

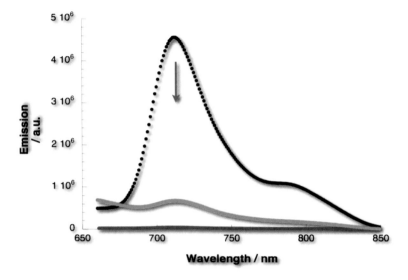

Fig. 9.43 Room temperature fluorescence spectra of the C_{60} reference **1** (*black*), **18a** (*red*) and **18b** (*orange*) in THF displaying the quenching of the fullerene emission in the triads

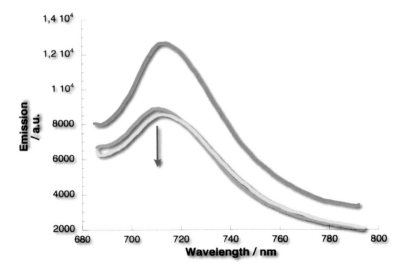

Fig. 9.44 Room temperature fluorescence spectra of **18b** in toluene (*dark orange*), THF (*light orange*) and benzonitrile (*yellow*) displaying the increasing quenching of the fullerene emission upon increasing solvent polarity

with short 387 nm (150 fs) and long 355 nm (5 ns) laser pulses, should be mentioned first. On the femtosecond time scale, the transient absorption spectra confirm the successful formation of the C_{60} singlet excited state with a maximum at 880 nm. At the wavelength of excitation both the oFL units and C_{60} absorb highly. Again, the singlet excited-state energy transfer from the linker to the fullerene moiety is established. In the corresponding C_{60}–oFL$_n$ and C_{60}–oFL$_n$–C_{60} conjugates, the formed singlet excited state intersystem crosses to form the triplet manifold on a time-scale of about 1.5 ns. In contrast to this finding, in the C_{60}–oFL$_n$–exTTF conjugates the singlet excited state features decay much faster than the mentioned 1.5 ns. Overall, the decay characteristics resemble the fluorescence experiments. In other words, the singlet excited state decay is faster in polar solvents (i.e. THF and benzonitrile) and slower upon increasing the oFL linker length. More importantly, at the expense of the singlet excited-state transient new transient features develop synchronously (Fig. 9.45). Significantly, one maximum is seen in the visible region at 680 nm and another maximum in the near infrared region at 1000 nm. We have already assigned these maxima to the one-electron oxidized exTTF radical cation (i.e. 680 nm) and the one-electron reduced C_{60} radical anion (i.e. 1000 nm) in Sect. 9.1.1. Hence, these transients confirm the intramolecular charge transfer reaction leading to $C_{60}^{\bullet-}$ and exTTF$^{\bullet+}$. Fitting the singlet excited-state decay and the radical ion pair formation kinetics (Fig. 9.45—lower part) yields the charge-separation rates (Table 9.5).

Since the radical ion pair states are stable on the time-scale of the femtosecond experiments, the charge-recombination rates were analyzed in complementary nanosecond experiments (Fig. 9.46). Therein, the decays of the $C_{60}^{\bullet-}$ and exTTF$^{\bullet+}$ features result in the refurbishment of the singlet ground state of **18a,b** lacking any detectable triplet features. The corresponding rate constants for the charge-recombination process are listed in Table 9.5.

The analysis of the charge-separation and charge-recombination rate constants enabled us to determine the β factor for the C_{60}–oFL$_n$–exTTF conjugates. Plotting the rates as a function of donor–acceptor distance R_{DA} results in linear relationships from which $\beta = 0.09 \, \text{Å}^{-1}$ emerged. Figure 9.47 represents the charge-separation and charge-recombination dynamics as a function of donor–acceptor distance.

9.2.1.2 Molecular Modeling

It was further possible to demonstrate the favorable charge-transfer interactions of the C_{60}–oFL$_n$–exTTF systems in a series of quantum chemical calculations and hypothetically expand this series to the trimer-based system, i.e. that bearing three oligofluorene units as linker. In order to compare the molecular-wire behavior of the oFL wires, we have used very similar methods as for the corresponding C_{60}–oPPE$_n$–exTTF systems (see Sect. 9.1.1).

For instance, inspecting the HOMO/LUMO energies of the building blocks by DFT methods (*B3LYP*/6−31*G**), i.e. exTTF, N-methylfulleropyrolidine **1**, the

9.2 oligo-Fluorene Molecular Wires

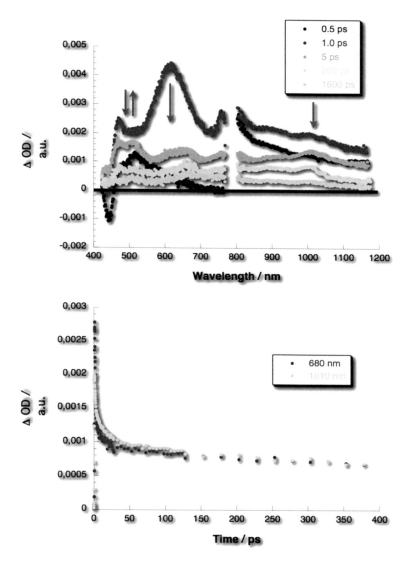

Fig. 9.45 *Upper part*—differential absorption spectra (visible and near-infrared) obtained upon femtosecond flash photolysis (387 nm–200 nJ) of **18b** in argon saturated THF with several time delays between 0 and 1600 ps at room temperature—illustrating the charge transfer. *Lower part*—time profiles at 680 nm ($exTTF^{\bullet+}$) and 1010 nm ($C_{60}^{\bullet-}$)—monitoring the charge-separation process

pristine oFL oligomers and the C_{60}–oFL dyads give results that resemble those found for the corresponding C_{60}–oPPE$_n$–exTTF triads. The HOMO/LUMO energies of exTTF in vacuo (-4.7 and -1.2 eV for HOMO and LUMO, respectively) perfectly match the energies of the oligomeric building blocks (-5.1 and -1.2 eV). Conclusively, the orbital overlap between the orbitals of donor and bridge is favored

Table 9.5 Charge-separation and charge-recombination dynamics as determined by femtosecond and nanosecond time-resolved spectroscopic measurements for **18a** and **18b** in THF

	R_{DA} [Å]	k_{CS} [s^{-1}]	k_{CR} [s^{-1}]
18a	16.7	8.9×10^9	7.2×10^5
18b	24.9	4.0×10^9	4.4×10^5

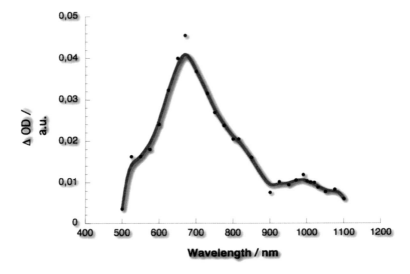

Fig. 9.46 Differential absorption spectrum (visible and near-infrared) obtained upon nanosecond flash photolysis (355 nm) of **18b** in nitrogen saturated THF (1.0×10^{-5} M) with a time delay of 100 ns at room temperature monitoring the formation of the *ex*TTF radical cation (680 nm) and the C$_{60}$ radical anion (1010 nm)

by the corresponding values of the orbital energies and the transfer of electrons onto the bridge is facilitated. Interestingly, the ionization potentials and electron affinities of the oligofluorenes remain invariant upon increasing the chain length. This has already been shown in the Molecular Modeling section of Sect. 8.1.2. On the other hand, attaching the fullerene moiety to the *o*FL chains lowers their LUMO energy to the LUMO level of pristine *N*-methylfulleropyrolidine **1**. Importantly, this occurs independently of the length of the oligofluorene chain (Fig. 9.48). Nonetheless, the HOMO energy remains unchanged in comparison to pristine fluorenes, reflecting the invariance of the oxidation potential of the oligofluorenes. Hence, this proves the electron-accepting features of C$_{60}$ and neglects electronic communication between the donor, bridge and acceptor in the ground state.

Figure 9.49 represents the HOMO/LUMO orbital-energy distribution in **18a,b** and the hypothetical trimer with three oligofluorene units as linker. Again, we clearly distinguish the HOMO and LUMO orbitals localized on the donor and acceptor, respectively, demonstrating the charge-transfer features in these triads.

9.2 oligo-Fluorene Molecular Wires

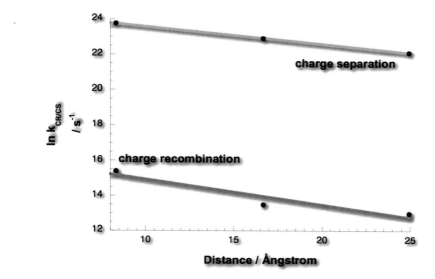

Fig. 9.47 Centre-to-centre distances (R_{CC}) dependence of charge-separation (ln k_{CS}) and charge-recombination (ln k_{CR}) rate constants in C_{60}–oFL$_n$–exTTF in nitrogen saturated THF at room temperature. The slope represents β

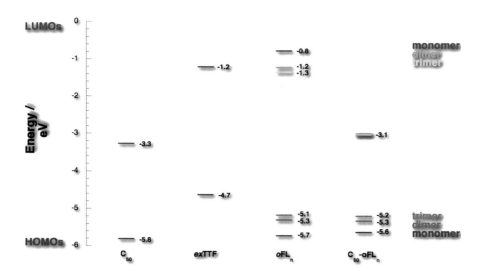

Fig. 9.48 HOMO/LUMO orbital energies as resulted from DFT optimizations of the building blocks of the triads. The HOMO energies are represented in *red* and LUMO energies in *green*

Additionally, electron-affinity calculations confirm these findings. Especially with respect to the oPPV and oPPE molecular wires, local affinity mappings as

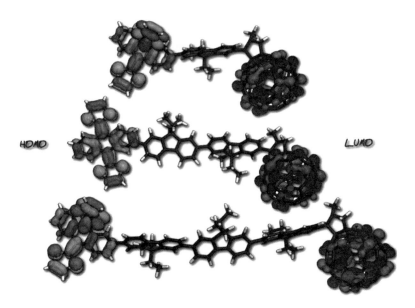

Fig. 9.49 HOMO/LUMO orbital schemes as resulted from DFT optimizations displaying the donor–acceptor character of the triads. The HOMOs are represented in *orange–blue* and LUMOs in *green–blue*

represented for the trimers of corresponding exTTF–WIRE–C_{60} triads in Fig. 9.50, place the charge-transfer properties of the oFLs in between the oPPVs and oPPEs. This is perfectly in line with the trend of the photophysically determined attenuation factors β, namely $\beta = 0.01$ Å^{-1} for oPPVs, $\beta = 0.09$ Å^{-1} for oFLs and $\beta = 0.21$ Å^{-1} for oPPEs. As suggested by Fig. 9.50, the electron affinity pathway in the oligofluorene based triads is more homogeneous than in exTTF–oPPE$_3$–C_{60} and less pronounced than in exTTF–oPPV$_3$–C_{60}. For the exTTF–oFL$_3$–C_{60}, the electron density (surface) is equally distributed throughout the whole molecule followed by a channel of high local electron affinity through the oligofluorene bridge, resulting in a maximum at the fullerene moiety.

As for the semi-empirical excited-state calculations from Sect. 9.1.1, we have performed CI calculations on the exTTF–oFL$_n$–C_{60} triads ($n = 1$–3). Unsurprisingly, the computations predict the HOMO to LUMO transition to be the major contribution to the charge-transfer reaction with a very high change of dipole moment, i.e. 80.8 Debye for the monomer, 117.7 Debye for the dimer and 160.85 Debye for the trimer. Visualization of the electrostatic potential of the charge-separated states confirms the electron donating character of the exTTF moiety and the electron accepting character of the C_{60} cage. In all triads, the positive charge was found to be localized on the exTTF and the negative one on the fullerene. Figure 9.51 shows the electrostatic potential of the charge-separated states of the exTTF–oFL$_n$–C_{60} triads.

9.2 oligo-Fluorene Molecular Wires

Fig. 9.50 Local electron affinity maps of the exTTF–oPPV$_3$–C$_{60}$, the exTTF–oFL$_3$–C$_{60}$ and the exTTF–oPPE$_3$–C$_{60}$ trimer (left to right) scaling high to low in *red* to *blue*—displaying the differences between the three different molecular-systems

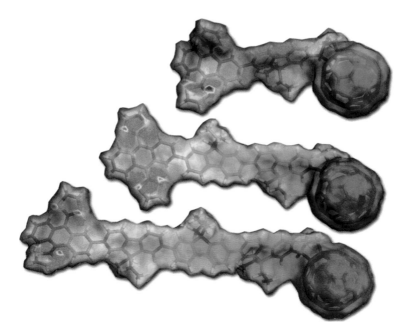

Fig. 9.51 Electrostatic potential as calculated by AM1 CIS for the charge-separated states of exTTF–oFL$_n$–C$_{60}$ triads ($n = 1$–3). Positive to negative: *red* to *blue*

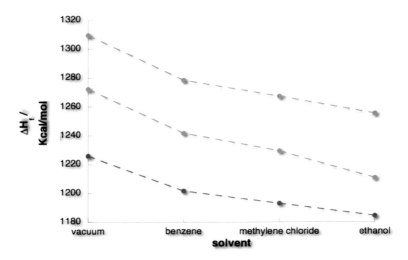

Fig. 9.52 Calculated heats of formation, ΔH_f, of the charge-separated states of exTTF–oFL$_n$–C$_{60}$ in vacuo and different solvents with increasing polarity; $n = 1$: *red*, $n = 2$: *orange*, $n = 3$: *pink*

Finally, we have examined the solvent dependence of the relative energies of the excited states. Herein, the single-point calculations on the relaxed structures of the ground and charge-transfer states in the simulated solvents benzene ($\epsilon = 2.3$), methylene chloride ($\epsilon = 8.9$) and ethanol ($\epsilon = 24.6$) should be mentioned. We have chosen these solvents due to their dielectric constant ϵ, which matches the solvent polarity of the solvents used in the photophysical investigation, i.e. toluene ($\epsilon = 2.4$), THF ($\epsilon = 7.6$) and benzonitrile ($\epsilon = 24.9$). Figure 9.52 shows the dependence of the calculated heats of formation, ΔH_f, of the charge-separated states on solvent polarity for the three oligomers of exTTF–oFL$_n$–C$_{60}$. In accordance with the experimental trends, in all systems the energies of the charge-separated states decrease with increasing solvent polarity due to the stabilization of the radical ion pair by polar solvent molecules. Thus, the calculations support the underlying hypothesis of the solvent dependence of the charge-transfer processes.

9.2.1.3 Summary

We have demonstrated that the donor-acceptor conjugates **18a,b** exhibit efficient charge-transfer processes upon photoexcitation over distances of more than 24 Å. The charge-transport mechanisms are comparable to those established for the corresponding exTTF–oPPV$_n$–C$_{60}$ and exTTF–oPPE$_n$–C$_{60}$ systems. In view of the molecular-wire behavior of the oFLs, we have shown that in fluorene-based oligomers the ability to conduct charges lies between that of oPPVs and that of oPPEs. A β of 0.09 Å^{-1} and the quantum chemical investigation confirm this conclusion. Calculations performed with the hypothetically designed trimer suggest that the charge-transfer features will prevail in oligomers of considerable length. It may indeed be an interesting issue to probe longer oFL-based conjugates and examine their charge-transfer properties especially in light of different change-transfer mechanisms. A particularly appealing espousal is that the oxidation potential of *oligo*-fluorenes remains unchanged upon changing the number of fluorene units. Electron transfer mediated by oFL bridges should be independent on the donor–acceptor distance in a series of donor–oFL–acceptor conjugates.

9.2.2 ZnP–oFL–C_{60} and Ferrocene–oFL–C_{60} Donor–Acceptor Conjugates

In the context of *oligo*-fluorene (oFL) molecular wires, we will conclude with an outlook that focuses on donor–acceptor conjugates bearing C_{60} electron-accepting units and zinc porphyrin (ZnP) or ferrocene (Fc) electron-donating moieties, respectively. Both, the donor and acceptor are covalently connected by oFL bridges of variable length.

Owing to the fact that the investigation of the photophysical behavior of these compounds has not yet been fully accomplished, we will limit the discussion to some interesting results. These already enable comparing the photophysics of the current systems with the one obtained in the previously described exTTF–oFL$_n$–C_{60} conjugates (see Sect. 9.2.1). In particular, the effects of the exchanging the exTTF by the corresponding ZnP and Fc electron donors will be worked out in this chapter. For that reason, we will supply the reader with some comparative photophysical characteristics.

9.2.2.1 ZnP–oFL$_n$–C_{60}

Particularly interesting in the ZnP–oFL$_n$–C_{60} conjugates is the comparison with the already discussed ZnP–oPPE$_n$–C_{60} (see Sect. 9.1.2). Let us recall that exchanging exTTF by ZnP has significant influence on the molecular-wire behavior of the oPPEs. The electron rich nature of ZnP, for instance, extends the π-conjugation by transferring electron density to the adjacent parts of the bridges. This, in turn, impacts the electron-injection process and therefore facilitates the charge-transfer processes. As a matter of fact, faster charge-separation rates and lower β values were established when compared to the corresponding exTTF conjugates. Enthrallingly, all systems lack any significant interaction between their building blocks in the ground state. In that sense, it is interesting to have a short look at the influence of the ZnP as donor on the molecular-wire behavior of the oFLs. For a schematic representation of the conjugates and their corresponding reference compounds, please consider Fig. 9.53.

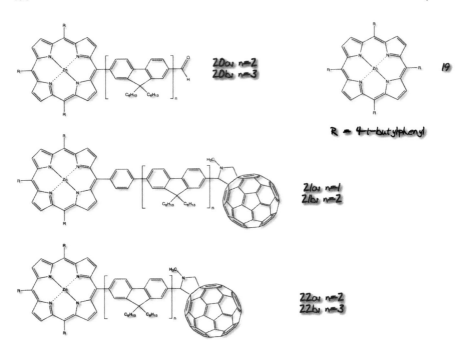

Fig. 9.53 Schematic representation of the ZnP–oFL$_n$–C$_{60}$ triads and their reference compounds

Let us first examine the ground-state absorption features of **21a,b** and **22a,b**. Figure 9.54 displays the absorption spectra of the four different oligomers. The absorption maxima of the building blocks are clearly distinguishable and resemble the absorption maxima of the references, i.e. the ZnP donors (Q-band: 425 nm; Soret bands: 550 and 590 nm), the oFL bridges (340 nm) and the C$_{60}$ acceptors (below 300 nm). Thus, it is safe to postulate that electronic interactions between their building blocks in the ground state are absent. However, it is important to point out that the *oligo*-fluorene absorption red shifts gradually between 310 and 390 nm. From the monomer to the trimer the maximum is shifted by more than 25 nm due to the immense extension of the π-conjugation in the bridge. Notably, the additional phenyl unit placed between oFL and ZnP in **21a,b** alters the π-system of the conjugates without interrupting the conjugation. The gradual shift of the maximum by approximately 8 nm when adding one fluorene (or phenylene) unit suggests that the π-system is uniformly extended over the whole conjugate. This is in line with previous results on ZnP–oPPE$_n$–C$_{60}$ conjugates. Insertion of ZnP leads to an increase of electron density at the donor–bridge junction and to full conjugation of the bridge π-system and the porphyrin π-electrons. Furthermore, the porphyrin absorption maximum does not depend on the length of the bridge, which, again, proves the lack of electronic interaction in the ground state.

Inspecting the emission features of the conjugates, we see the already known characteristics of a successful electron-transfer reaction. Regardless of the

9.2 oligo-Fluorene Molecular Wires

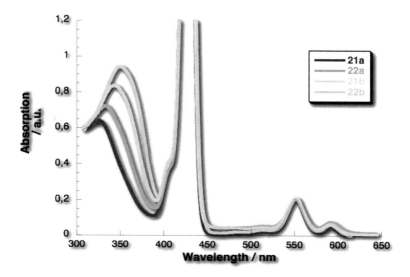

Fig. 9.54 UV–visible absorption spectra of **21a,b** (*green* and *yellow*) and **22a,b** (*light green* and *pink*) in toluene solution

excitation wavelength, i.e. 345 nm for oFL excitation and 550 for ZnP excitation, significant quenching of the porphyrin emission is present in all systems. The quenching is once more dependent on the length of the oFL linker and on the solvent polarity. As expected, the oFL fluorescence at 430–450 nm is appreciably quenched due to efficient singlet excited-state energy transfer from the photoexcited oligomer unit to C_{60}. Following the established trends, the quenching of the porphyrin emission indicates the involvement of a charge transfer process. Specifically, the ZnP porphyrin fluorescence quenching in **21a,b** and **22a,b** depends on the length of the bridge (Fig. 9.55). In particular, strongest quenching is found in the monomer **21a**, with a quantum yield in toluene of 8.0×10^{-4} corresponding to a factor of 50, whereas the quenching in the trimer **22b**, which shows a quantum yield of 2.0×10^{-2}, amounts only to a factor of 2. Interestingly, this occurs regardless of the excitation wavelength, which infers that the presence of C_{60} enhances the fluorescence deactivation.

Inspecting the solvent dependence of the quenching indicates electron-transfer processes in all oligomers of **21** and **22**. With increasing solvent polarity from toluene ($\epsilon = 2.4$) to THF ($\epsilon = 7.6$) and benzonitrile ($\epsilon = 24.9$), the quenching of the porphyrin emission increases (Fig. 9.56).

These findings have been corroborated by fluorescence lifetime measurements (Table 9.6). Notably, the ZnP emission lifetime in the conjugates tends to be much shorter in comparison with the ZnP reference compounds. With increasing solvent polarity the fluorescence deactivation is further accelerated, i.e. the emission lifetimes decrease which is in line with the gradual deintensification of the fluorescence. Once more, such behavior is indicative for a deactivation channel where

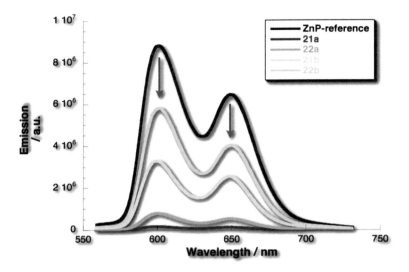

Fig. 9.55 Room temperature fluorescence spectra of the ZnP-reference (*black*), **21a,b** (*green* and *yellow*) and **22a,b** (*light green* and *pink*) in toluene displaying the quenching of the porphyrin emission in the triads upon 345 nm excitation

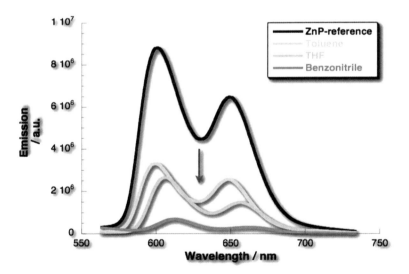

Fig. 9.56 Room temperature fluorescence spectra of the ZnP-reference (*black*) and **21b** in toluene, THF and benzonitrile displaying the solvent dependence of the quenching of the porphyrin emission upon 345 nm excitation

9.2 oligo-Fluorene Molecular Wires

Table 9.6 Fluorescence lifetimes and charge-separation rates of the triads and their reference compounds

	τ_1 [ns]	k_{CS} [s^{-1}]	R_{CC} [Å]
19	Toluene: 1.46 THF: 1.40 Benzonitrile: 1.36	–	–
20a	Toluene: 1.31 THF: 1.31 Benzonitrile: 1.19	–	–
20b	Toluene: 1.36 THF: 1.33 Benzonitrile: 1.35	–	–
21a	Toluene: 0.26 THF: 0.19 Benzonitrile: 0.12	1.48×10^9	21.08
21b	Toluene: 0.34 THF: 0.23 Benzonitrile: 0.20	7.69×10^8	25.06
22a	Toluene: 0.86 THF: 0.69 Benzonitrile: 0.59	6.05×10^8	28.85
22b	Toluene: 1.16 THF: 0.82 Benzonitrile: 0.68	4.46×10^8	32.39

an intramolecular electron transfer between the porphyrin donor and the C$_{60}$ acceptor is operative. The charge-separation rates are between 10^8 and 10^9 s^{-1}. To sum up, the deactivation of the porphyrin fluorescence at 600 nm occurs on a very fast time-scale, which is due to the successful formation of the radical ion pair stabilized by solvents with higher polarity.

Finally, transient absorption measurements were deemed necessary to confirm the photoproducts in **21a,b** and **22a,b**. Due to overlapping absorptions of C$_{60}$, oFL and ZnP, which would impede a clear analysis, we have focused first on the selective excitation of ZnP. To this end, transient absorption spectra of the reference compounds (**19** and **20a,b**) reveal the instantaneous formation of the ZnP singlet excited state with maxima at 460 and 800 nm and minima at 565 and 605 nm. Furthermore, an isosbestic point at 500 nm as it develops on a time scale of 3000 ps reflects the intersystem crossing process at which end the triplet excited state of ZnP stands. The latter includes maxima at 530, 580 and 640 nm (Fig. 9.57a). Equally important is the fact that the decay of the singlet excited state matches the formation of the triplet excited state kinetics (Fig. 9.57b).

In comparison to the reference compounds, the transient absorption features of **21a,b** and **22a,b** disclose a rather fast deactivation of the ZnP singlet-singlet absorption. However, at the conclusion of the singlet decay the recorded transient spectra lack any triplet excited state signatures. Instead, the singlet–singlet

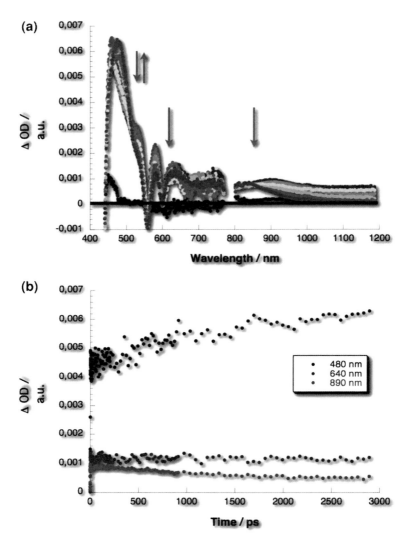

Fig. 9.57 a Differential absorption spectra (visible and near-infrared) obtained upon femtosecond flash photolysis (420 nm–150 nJ) of **20a** in argon saturated THF with several time delays between 0 and 3000 ps at room temperature—illustrating the instantaneous formation of the singlet-excited state of ZnP and the intersystem crossing. **b** Time profiles at 480, 640 and 890 nm reflecting the slow intersystem-crossing dynamics

deactivation results in new features with maxima at 480 and between 600 and 700 nm, as well as in the near-infrared region, i.e. 1000 nm (Fig. 9.58a). As we have already shown throughout the investigation of the corresponding ZnP–oPPE$_n$–C$_{60}$ conjugates, these features are attributed to the one-electron oxidized π-radical cation of ZnP (ZnP$^{\bullet+}$) and the one-electron reduced anion of C$_{60}$ (C$_{60}^{\bullet-}$).

9.2 *oligo*-Fluorene Molecular Wires

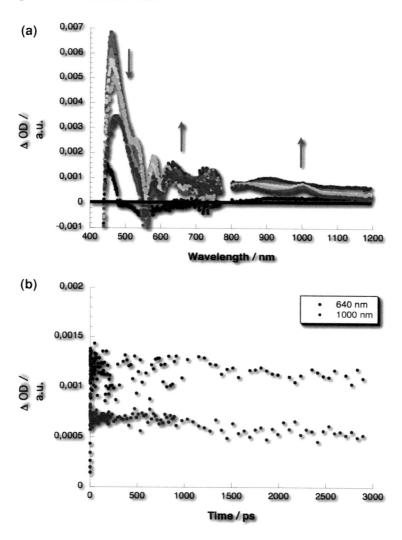

Fig. 9.58 **a** Differential absorption spectra (visible and near-infrared) obtained upon femtosecond flash photolysis (420 nm–150 nJ) of **22a** in argon saturated THF with several time delays between 0 and 3000 ps at room temperature. **b** Time profiles at 640 and 1000 nm reflecting the charge-separation dynamics

Summarizing, transient absorption measurements confirm the successful formation of the ZnP$^{\bullet+}$–oFL$_n$–C$_{60}^{\bullet-}$ radical ion pair upon photoexcitation.

With the charge-separation dynamics-based on the decay rates resulting from the time profiles of the C$_{60}$ π-radical anion and ZnP cation (Fig. 9.58b) - in hands we determined β as 0.09 Å^{-1} (Fig. 9.59). This value is in excellent agreement with that determined in the corresponding C$_{60}$–oFL$_n$–exTTF conjugates. At this point,

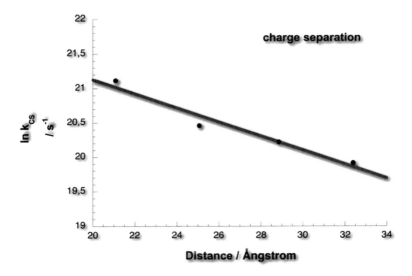

Fig. 9.59 Centre-to-centre distances (R_{CC}) dependence of the charge-separation (ln k_{CS}) rate constants in C_{60}–oFL$_n$–ZnP in nitrogen saturated THF at room temperature. The slope represents β

we recognize that exchanging the donor in oFL based donor–acceptor conjugates bearing C_{60} electron acceptors does not influence the transport properties of oFL. Nevertheless, to further corroborate our initial findings complementary transient absorption measurements (i.e. nanosecond) are necessary to determine the charge-recombination rate constants.

9.2.2.2 Fc–oFL$_n$–C$_{60}$

Similar to our assays with the ZnP–oFL$_n$–C$_{60}$ conjugates, we have employed a number of photophysical measurements with complementary Fc–oFL$_n$–C$_{60}$ systems ($n = 1, 2$). Herein, the ZnP donor has been substituted by a ferrocene moiety (Fig. 9.60). In view of the the *oligo*-fluorenes in ZnP–oFL$_n$–C$_{60}$ conjugates which lead to an attenuation factor of 0.09 Å^{-1}, it was of particular interest to probe the influence of yet another donor, namely ferrocene, on β. Notable is the perfect match of the attenuation factor with that in exTTF–oFL$_n$–C$_{60}$ (see Sect. 9.2.1). This leads us to postulate that (oFL)$_n$ may exhibit a molecular wire behavior that is truly independent on the donor moieties. For that reason, probing the charge-transfer properties of **24a,b** provided another opportunity to prove this hypothesis and examine the influence of the ferrocene donor on the electron-transfer behavior of (oFL)$_n$.

We have already gathered testimony for the lack of electronic interaction between the individual building blocks in ZnP–oFL$_n$–C$_{60}$. In line with the expectation that Fc should not change this trend, the ground-state absorption

9.2 oligo-Fluorene Molecular Wires

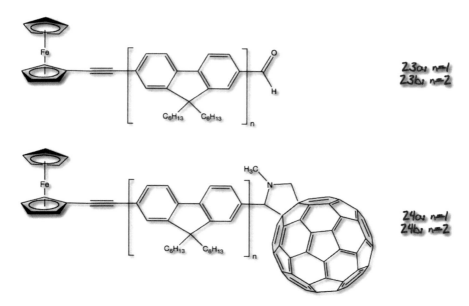

Fig. 9.60 Schematic representation of the Fc–oFL$_n$–C$_{60}$ triads and their reference compounds

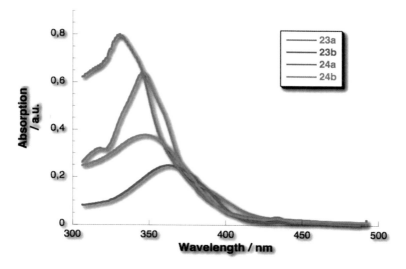

Fig. 9.61 UV–visible absorption spectra of **23a,b** (*pink* and *brown*) and **24a,b** (*orange* and *green*) in toluene solution

studies of **24a,b** unveil just the same superimposing features of the building blocks. Figure 9.61 summarizes the absorption spectra of **23a,b** and **24a,b** in toluene. The *oligo*-Fluorene absorption dominates the spectral features in the visible range (300–400 nm). A length dependence extends the π-conjugation in the dimers—compared to the monomers—and shifts the absorption to the red. Distinct fullerene absorption features are found at 433 nm. These do not depend on the length of the oFL bridge. Interestingly, the maxima of the oFL absorption in the references **23a,b** are red-shifted by around 15 nm in comparison to the conjugates. Considering the terminal aldehyde groups present in **23a,b**, their π-electrons contribute to the overall π-conjugation of the *oligo*-fluorene moieties. Replacing now the aldehyde groups by N-methyl-fulleropyrolidines limits the conjugation to the $(o\text{FL})_n$ core. Attributable to very low extinction coefficients of Fc, they are not discernible. Summarizing, our UV/Vis absorption studies infer the lack of electronic interactions between the single components in in Fc–oFL$_n$–C$_{60}$, namely ferrocene, oFL bridge and C$_{60}$, in the ground-state.

The outcome of the steady-state emission studies is perfectly in line with the results obtained throughout the study of the corresponding donor–oFL$_n$–C$_{60}$ systems presented above. For instance, upon excitation of the oFL building blocks, i.e. 345 nm, quantitative quenching of the *oligo*-fluorene emission is present in both, monomer **24a** and dimer **24b** (Fig. 9.62a). Consequently, the oFL singlet excited state is instantaneously deactivated by intramolecular energy-transduction processes - product is the energetically lower lying C$_{60}$ singlet excited state. Further insight into the fluorescence deactivation came from inspecting the 710 nm C$_{60}$ fluorescence (Fig. 9.62b). Here, significant quenching of the C$_{60}$ fluorescence was evidenced in **24a** and **24b**. This indicates that the *oligo*-fluorene singlet excited-state deactivation is governed by subsequent charge transfer which evolves between photoexcited C$_{60}$ and ferrocene.

Additionally, the quenching depends on the oFL length and on solvent polarity. This is reflected in the fluorescence quantum yields (Table 9.7). Importantly, with increasing donor-acceptor distance the quantum yields increase due to slower charge-separation dynamics. On the other hand, increasing the solvent polarity causes the quantum yields to decrease due to better stabilization of the radical ion pair. The fluorescence lifetimes (Table 9.7) confirm the trend of acceleration of charge-separation by solvents of higher polarity indicating again that the deactivation of the C$_{60}$ fluorescence occurs via an intramolecular electron-transfer reaction and results in the formation of the radical ion pair state.

Further evidence for the proposed charge-transfer processes in **24a,b** was found in complementary transient absorption measurements on the femto- to picosecond time scale. At the chosen excitation wavelength of 387 nm, oFL and C$_{60}$ were excited in a similar ratio. Considering the reference compounds **23a,b**, photoexcitation generates instantaneously meta-stable singlet excited state transients. Characteristics of the latter are a ground-state bleaching in the 400–450 nm region and transient absorptions in the 600–1200 nm region (Fig. 9.63a). The formation is accomplished on a time-scale of less than 3 ps (Fig. 9.63b). Furthermore, we were able to establish that the transient maxima depend on the oFL length: 615 nm (**23a**) and 630 nm (**23b**). In all oligomers, the product of the singlet decay is the corresponding oFL triplet excited-state with new characteristics in the 600–800 nm region.

9.2 oligo-Fluorene Molecular Wires

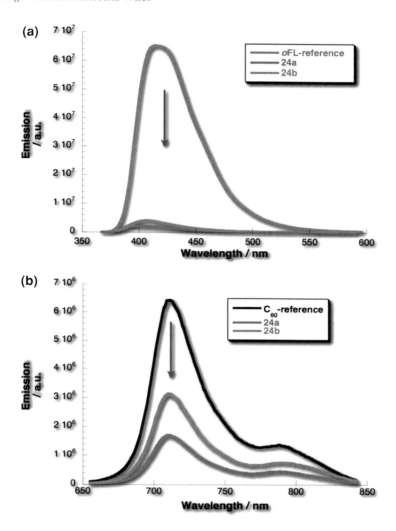

Fig. 9.62 a Fluorescence spectra (345 nm excitation) of the oFL reference (*pink* spectrum), **24a** (*orange* spectrum) and **24b** (*brown* spectrum) in toluene representing the quenching of the oFL emission. b Fluorescence spectra of the C_{60} reference (*black* spectrum), **24a** (*orange* spectrum) and **24b** (*brown* spectrum) in toluene representing the quenching of the C_{60} emission

In contrast to the references, **23a,b**, the singlet-excited state deactivation in the Fc–oFL_n–C_{60} conjugates, **24a,b**, occurs with rate constants of around 10^{10} s^{-1} (Fig. 9.64). The values are in good agreement with the observed quantitative quenching of the oFL fluorescence (Fig. 9.62a). At the conclusion of the singlet-excited state decay, two important transient maxima resemble the successful formation of the radical ion pair state, namely a weak shoulder of the transient

Table 9.7 Selected spectroscopic data for the Fc–oFL$_n$–C$_{60}$ systems and their reference compounds

	τ_1 [ns]	ϕ_f	k_{CS} [s^{-1}]	R_{CC} [Å]
24a	Toluene: 0.88	1.58×10^{-4}		
	THF: 0.75	4.28×10^{-5}	4.29×10^9	16.77
	Benzonitrile: 0.69	2.97×10^{-5}		
24b	Toluene: 1.35	2.90×10^{-4}		
	THF: 1.22	1.99×10^{-4}	6.57×10^8	24.88
	Benzonitrile: 0.93	1.11×10^{-4}		

absorption of the one-electron oxidized ferrocenium at 630 nm [16] and the one-electron reduced C$_{60}$ at 1000 nm (Fig. 9.64). Thus, the singlet-excited state deactivation in the conjugates results in the generation of the Fc$^{\bullet +}$–oFL$_n$–C$_{60}^{\bullet -}$ radical ion pair. Importantly, the formation of the radical ion pair in **24a** occurs with kinetics from which a rate constant for the charge-separation was determined that amounted to 4.3×10^9 s^{-1} in THF. On the other hand, charge-separation in the dimer **24b** is slower by one order of magnitude, i.e. with a rate of 6.6×10^8 s^{-1} in THF.

The charge-separation dynamics as deduced from the decays of the Fc radical cation and the C$_{60}$ radical anion characteristics as a function of donor–acceptor distance (Fig. 9.65) yielded a linear relationship. From the slope β was determined as 0.19 Å^{-1}. Such a value is twice as high as the values derived for C$_{60}$–oFL$_n$–exTTF and C$_{60}$–oFL$_n$–ZnP. However, in order to further corroborate this value, it is necessary to examine the charge-recombination processes on the nanosecond time-scale and to further expand this series to the trimer, C$_{60}$–oFL$_3$–Fc, tetramer, C$_{60}$–oFL$_4$–Fc, etc.

9.2.2.3 Summary

On account of the thoroughly characterized C$_{60}$–oFL$_n$–exTTF systems in Sect. 9.2.1, the first insight into corresponding conjugates bearing ZnP and Fc donors implies that oFLs manifest charge-transfer behavior, which is at first glance independent on the donor moiety. In C$_{60}$–oFL$_n$–ZnP, β matches exactly that measured in C$_{60}$–oFL$_n$–exTTF, whereas in C$_{60}$–oFL$_n$–Fc the wire-like behavior turned out to be less efficient. Several factors might be responsible for this trend. Most importantly, Fc does not conjugate with oFL as the corresponding exTTF and ZnP. Furthermore, in the C$_{60}$–oFL$_n$–Fc system, the Fc donor and the oFL moieties are connected through an additional ethynylene linker. Triple bonds have shorter lengths than the corresponding double bonds of the oFL π-system and, in turn, this prevents a full conjugation between the building blocks, i.e. the donor, bridge and acceptor. Similarly, in the C$_{60}$–oPPE$_n$–exTTF conjugates the triple bonds in the oligo(para-phenyleneethynylene) linkers disrupt the π-conjugation due to altering

9.2 oligo-Fluorene Molecular Wires

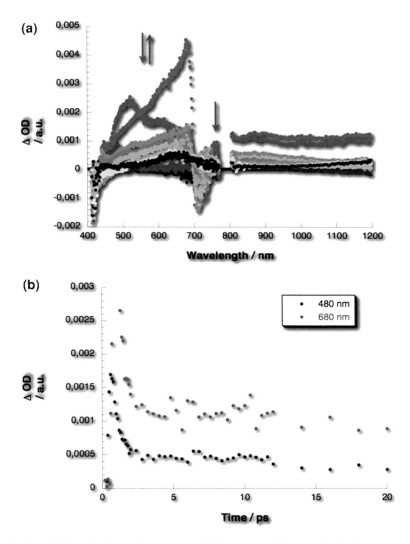

Fig. 9.63 **a** Differential absorption spectra (visible and near-infrared) obtained upon femtosecond flash photolysis (387 nm–150 nJ) of **23b** in argon saturated THF with several time delays between 0 and 3000 ps at room temperature—illustrating the instantaneous formation of the singlet-excited state and the intersystem crossing. **b** Time profiles at 480 and 680 nm reflecting the rapid formation of the oFL singlet excited state

bond lengths in an interplay with the double bonds of the phenyl rings. When considering charge-separation processes, porphyrin donor moieties are undoubtedly more advantageous in comparison to ferrocenes: Larger absorption cross sections and energetically more accessible singlet–singlet transitions are the most notable proponents. Furthermore, heavy atom effects and high spin–orbit coupling

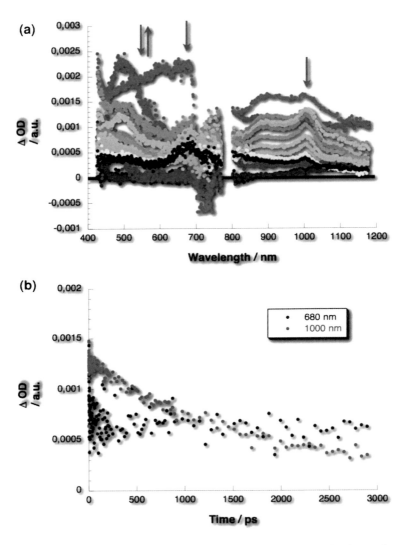

Fig. 9.64 a Differential absorption spectra (visible and near-infrared) obtained upon femtosecond flash photolysis (387 nm–150 nJ) of **24b** in argon saturated THF with several time delays between 0 and 3000 ps at room temperature—illustrating the formation of the radical ion pair state. **b** Time profiles at 680 and 1000 nm reflecting the formation of the radical ion pair

of the iron atom should be mentioned. Facile deactivation of excited states due to significant overlap between the metal d-orbitals and the p-orbitals of the adjacent ethynylene linker emerges as a competitive scenario. Certainly, the evaluation of the energies of the building blocks, e.g. HOMO and LUMO levels, will provide further insight into the interactions between the modular components of the

9.2 *oligo*-Fluorene Molecular Wires

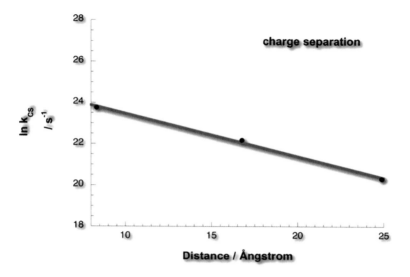

Fig. 9.65 Centre-to-centre distances (R_{CC}) dependence of the charge-separation (ln k_{CS}) rate constants in C_{60}–oFL_n–Fc in nitrogen saturated THF at room temperature. The slope represents β

conjugates. Hence, electrochemical and quantum chemical examinations are required to fully understand the energetic relationship between the building blocks. This will help to work out the difference between the three donors (i.e. *ex*TTF, ZnP and Fc) and their influence on the charge-transfer properties of the *o*FLs.

References

1. Giacalone F, Segura JL, Martin N, Guldi DM (2004) J Am Chem Soc 126:5340
2. Torres G, Gialcone F, Segura JL, Martín N, Guldi DM (2005) Chem–Eur J 11:1267
3. Giacalone F, Segura JL, Martín, Ramey J, Guldi DM (2005) Chem–Eur J 11:4819
4. Guldi DM, Swartz A, Luo Ch, Gomez R, Segura JL, Martín N (2002) J Am Chem Soc 124:10875
5. Redmore NP, Rubstov IV, Therien MJ (2003) J Am Chem Soc 125:8769
6. Dong T-Y, Chang S-W, Kin S-F, Lin M-C, Wen Y-S, Lee L (2006) Organometallics 25:2018
7. Pourtis G, Beljonne D, Cornil J, Ratner MA, BrØdas JL (2002) J Am Chem Soc 124:4436
8. Wielopolski M, Atienza-Castellanos C, Clark T, Guldi DM, Martín N (2008) Chem–Eur J 14:6379
9. Winget P, Horn AHC, Selçuki C, Martin B, Clark T (2003) J Mol Mod 9:408
10. Halgren TA, Lipscomb WN (1977) Chem Phys Lett 49:225
11. Lembo A, Tagliatesta P, Guldi DM, Wielopolski M, Nuccetelli M (2008) J Phys Chem A submitted
12. Stewart JJP (1989) J Comput Chem 10:209
13. Weller Z (1982) Phys Chem Chem Phys 133:93
14. Gayathri SS, Wielopolski M, Pérez EM, Fernández G, Sánchez L, Viruela R, Ortí E, Guldi DM, Martín N (2008) Angew Chem Int Ed 48:815

15. Atienza-Castellanos C, Wielopolski M, Guldi DM, Pol Cvd, Bryce MR, Filippone S, Martín N (2007) Chem Commun (Cambridge, U. K.) 48:5164
16. Faraggi M, Weinraub D, Broitman F, DeFelippis MR, Klapper MH (1988) Radiat Phys Chem 32:293

Chapter 10
Conclusions and Outlook

Based on the results obtained throughout the presented investigation on molecule-assisted transport of charge and energy mediated by organic wire-like structures, fully conjugated organic aromatic molecular wires turned out to be the best candidates for introduction into new electronic devices as replacements for aluminium or copper wiring as presently utilized in logic and memory devices.

Specifically, energy- and charge-transfer properties of several different molecular-wire systems have been studied within the framework of photoinduced charge separation and solar-energy conversion. Up front, the conductance behavior of wire-like molecules was of particular interest. Such features have been carefully examined in view of possible applications in the fields of molecular electronics and/or photovoltaic devices. Among the tested systems, π-conjugation played a crucial role.

To be specific, we have focused on four different π-conjugated systems, namely (i) *oligo*(*para*-phenylenevinylene)s (*o*PPVs), (ii) *oligo*(*para*-phenyleneethynylene)s (*o*PPEs), (iii) *oligo*(*meta*-phenyleneethynylene)s (*o*MPEs) and (iv) *oligo*-Fluorenes (*o*FLs). Common to all these structures is an extended π-conjugation throughout the whole building blocks. Their length is easily modified by simple synthetic methods. In our studies, the wire-like molecules of varying length were covalently implemented into a variety of different supramolecular assemblies, i.e. acceptor–acceptor or donor–acceptor conjugates. In all cases, C_{60} fullerenes have been utilized as acceptor moieties due to their outstanding electron and energy transfer properties. The donor units have been varied—ranging from *ex*TTF and porphyrins (H_2P and ZnP) to ferrocenes (Fc). Considering the electron–donor–acceptor character of the conjugates, the subject matter of concern was the examination of photoinduced energy- and charge-transfer processes and the influence of molecular structure on those. For that reason, a variety of photophysical measurement techniques combined with quantum mechanical calculations have been employed. This helped in analyzing the molecule-assisted transport of charges and energy. The outcome of this scrutiny was assessing and categorizing the conduction properties as a function of chemical structure.

In the first part of this thesis, energy-transfer processes were investigated with oPPE and oFL containing conjugates, i.e. C_{60}–oPPE$_n$–C_{60} and C_{60}–oFL$_n$–C_{60}. We have demonstrated that the conjugated oPPE and oFL absorb light in a wide range of the visible region of the solar spectrum and are capable of directive transduction of excited state energy to, for example, C_{60}. In all cases, the unidirectional energy transfer turned out to be nearly quantitative and virtually independent on the length of the implemented oPPE or oFL. Importantly, such energy-transfer processes occur although no significant electronic interactions between the modular units are present in the ground state. Notably, oFL exhibit a length-independent oxidation potential. In other words, the energy of the HOMO/LUMO orbitals is practically invariant from the oligomer length. Such properties imply perfect conditions for efficient superexchange coupling between donor and acceptor even over large distances. Finally, we have concluded the investigation on energy-transfer processes in donor–acceptor system based on non-covalent hydrogen-bonding interactions. These assays serve as a confirmation of the energy accepting features of the fullerenes. Thereby, a novel concept of efficient through-space energy transfer has been demonstrated. In particular, structural preconditions have been formulated to the set the stage for efficient photoinduced energy transduction. Important factors include appropriate light-harvesting units, the tunability of the absorption cross-section and π-conjugation. Specifically, it has been shown that aromatic porphyrin chromophores act as versatile antenna systems for transmitting excited-state energy to noncovalently associated fullerene moieties upon photoexcitation. Most interestingly, the mediation of singlet excited-state energy in these systems was driven by noncovalent hydrogen-bonding interactions.

In view of the aforementioned energy-transfer processes and efficient light-harvesting properties, π-conjugated molecular wires have been examined with respect to their charge-transfer properties in the second part of this thesis. Undoubtedly, the foregoing analysis of excited state energy transfer was essential to understand photoinduced charge transfer in complementary donor–wire–acceptor conjugates. In other words, only in case of efficient light-harvesting and energy-transfer features the corresponding wire-like molecules qualify as good mediators for electron-transfer reactions in donor–wire–acceptor conjugates. As the mentioned molecules seem to meet these requirements, we turned to intramolecular electron transfer along oPPE, oMPE and oFL chains in several donor–acceptor conjugates. ExTTF, porphyrins and Fc act as electron donors and fullerenes as electron acceptors. In particular, the investigation demonstrated the importance of energy matching between the donor and bridge components for achieving true molecular-wire behavior. Furthermore, the competition between direct superexchange and two-step "bridge-mediated" charge-transfer processes has been elaborated. Particular emphasis was placed on the nature and length of the conjugated bridges. The systematic study of the different donor–bridge–acceptor compounds and their electronic properties was carried out by means of photophysical and quantum chemical methods.

At first, the focus was on charge-transport properties in oPPE containing donor–acceptor systems. To understand and to compare the experimental and theoretical outcome of the investigation we have taken into account the results obtained with complementary oPPV systems. Both, the exTTF–oPPE$_n$–C$_{60}$ and the corresponding H$_2$P/ZnP–oPPE$_n$–C$_{60}$ systems gave rise to charge-transfer interactions upon photoexcitation. Solvent dependent steady-state and time-resolved spectroscopic measurements revealed the formation of nanosecond stable radical ion pairs, namely exTTF$^{•+}$–oPPE$_n$–C$_{60}^{•-}$ and H$_2$P$^{•+}$/ZnP$^{•+}$–oPPE$_n$–C$_{60}^{•-}$. Furthermore, charge-separation and charge-recombination dynamics disclosed a fairly shallow dependence of charge-transfer rates on donor–acceptor distances yielding attenuation factors (β) as low as 0.20 and 0.11 Å^{-1} for exTTF–oPPE$_n$–C$_{60}$ and H$_2$P/ZnP–oPPE$_n$–C$_{60}$, respectively. The difference in attenuation factors was attributed to a better conjugation between donor and bridge in the case of porphyrin donors. It is likely that the higher π-electron density of the porphyrins is responsible for this trend. In line with the experimental observations, quantum mechanical calculations revealed a hole-transfer mechanism in the corresponding triads and fairly good orbital overlap between the building blocks. Additionally, calculations corroborated the improved electron-donating properties of the porphyrins relative to the exTTF moieties. Importantly, the differences in the attenuation factors, which are more than one order of magnitude from these obtained in complementary exTTF–oPPV$_n$–C$_{60}$ systems, were assigned to alternating bond lengths in oPPEs. In comparison to the oPPV molecular wires, the triple bonds in oPPE oligomers are shorter than the corresponding double bonds. Hence, in connection with their polarizing character, the π-conjugation in oPPEs is not as homogeneous as in oPPVs, which strongly impacts the efficiency of the charge-transfer processes.

Additionally, the effects of a different connectivity pattern on the conduction properties have been studied upon exchanging the oPPE molecular bridges by the corresponding *meta*-connected oligomers, i.e. oMPEs. In that context, a series of exTTF–oMPE$_n$–C$_{60}$ systems have been investigated by means of spectroscopic methods. It was found that the difference in electronic properties between the *para*- and *meta*-connected systems is mainly governed by the more compact structure of the latter. Furthermore, the *meta*-connectivity pattern comes along with a lack of π-conjugation, which, in turn, impacts the charge-transfer features, that is, weakening. As a matter of fact, no meaningful distance dependence was present in electron-transfer reactions. In fact, only the monomer of exTTF–oMPE$_n$–C$_{60}$ ($n = 1$) revealed a charge-transfer behavior that is driven by through-bond interactions. When directing our attention to the remaining oligomers, the charge-separation rates turned out to be independent on the length of the oMPE linker. However, the formation of stable radical ion pairs was proven in all exTTF–oMPE$_n$–C$_{60}$ systems. Hence, such findings indicate that charge transfer in exTTF–oMPE$_n$–C$_{60}$ occurs rather through space than through bond (i.e. oMPE$_n$). Accordingly, this is in line with the lack of π-conjugation in oMPEs. In other words, the formation of the radical-ion-pair state in the monomer is most likely due to very short distances between the donors and acceptors resulting from the *meta*-connectivity pattern. Such features have been only observed at low concentrations. In stark contrast, upon increasing the concentration novel charge-transfer features develop. An extensive analysis of the concentration-dependent charge-transfer properties disclosed interesting aggregation phenomena. Specifically, due to π–π-stacking interactions between the aromatic systems of exTTF and C_{60} complex formation between individual exTTF–oMPE$_n$–C_{60} molecules was observed. The so obtained $(ex$TTF–oMPE$_n$–$C_{60})_x$ aggregates promoted ground-state charge-transfer interactions. These were discernable in steady-state and time-resolved spectroscopic measurements. In particular, partial ground-state charge transfer was assumed to result from a shift of electron density from exTTF to C_{60}. Furthermore, the lifetimes of the charge-separated states tend to increase in comparison with the "monomeric" form, which might be due to a delocalization of the charge in the intracomplex hybrids.

Issuing from the findings obtained throughout the investigation of oPPE and oMPE molecular wires, similar analyses of donor–bridge–acceptor conjugates bearing oFLs have been carried out to evaluate the distance dependence of charge transfer. Several donor moieties were implemented reaching from exTTF and ZnP to ferrocene. This meant to elaborate the impact of their chemical nature on the conjugation between donor and bridge.

At first, the charge-transfer properties in exTTF–oFL$_n$–C_{60} systems have been elaborated in theory and experiment. In a similar fashion to the aforementioned exTTF–oPPE$_n$–C_{60} system, the formation of radical ion pairs was manifested by means of transient spectroscopy. The exTTF–oFL$_n$–C_{60} conjugates exhibit efficient charge-transfer processes over distances of more than 24 Å. Interestingly, the determined attenuation factor of 0.09 Å^{-1} suggests that the molecular-wire behavior of oFLs is between that of oPPVs and oPPEs. These findings have been corroborated by quantum mechanical calculations, which revealed a fairly homogeneous distribution of π-electron aromaticity throughout the whole molecules. An implicit factor is that, the exTTF donor orbitals do not conjugate significantly into the bridge. The invariant (i.e. as a function of length) oxidation potential of the oFL and their energetically high-lying LUMO orbitals are accountable for these findings and impede strong electronic coupling between donor and acceptor. Nevertheless, besides efficient π-conjugation, the strong electronic coupling is responsible for the very low attenuation factors in the corresponding exTTF–oPPV$_n$–C_{60} systems. On the other hand, the lack of bond-length alteration in oFLs improves the conduction behavior relative to oPPEs. In fact, a homogeneous distribution of electron density is seen.

An intriguing question to be addressed was the impact that the donor moiety exhibits on the charge transfer/charge transport features in oFLs. The difference seen in exTTF–oFL$_n$–C_{60} and ZnP–oFL$_n$–C_{60} provides incentives for this comparison. In $(o$FL$)_n$, placing exTTF as electron donor led to a β value of 0.09 Å^{-1}. Replacing exTTF by ZnP was the obvious choice. In fact, the photophysical outcome suggested rather donor-independent charge transfer and charge transport in the corresponding ZnP–oFL$_n$–C_{60} systems. In particular, charge-separation was proven by means of steady-state and time-resolved spectroscopic issues. Measurements in solvents of different polarity implied successful formation of radical-ion-pair states. Large charge-separation rate constants reflect a full conjugation involving the aromatic system of oFL and the porphyrin donor. Most importantly, the charge-separation rates exhibited a linear relationship to the donor–acceptor distance affording an attenuation factor, which exactly matches the β value established in the exTTF–oFL$_n$–C$_{60}$ conjugates. Hence, we hypothesize that the conductance in fully conjugated oFLs is independent on the donor moiety. Such observation is perfectly in line with the oxidation potentials as they were determined in conjunction with the energy transfer features in C$_{60}$–oFL$_n$–C$_{60}$ conjugates.

In contrast, the electron-transfer properties in Fc–oFL$_n$–C$_{60}$ donor–bridge–acceptor conjugates turned out to be less efficient. Nevertheless, the formation of charge-separated ion pairs was corroborated by various photophysical measurements in different solvents. Slower charge transfer reactions, when compared to exTTF–oFL$_n$–C$_{60}$ and ZnP–oFL$_n$–C$_{60}$, point to a lack of homogeneous conjugation in Fc–oFL$_n$–C$_{60}$. A β value of 0.19 Å^{-1} results from the corresponding distance dependence. A closer analysis suggests that 0.19 Å^{-1} approaches the values that are characteristic for (oPPE)$_n$ systems. To fully rationalize these findings further analysis deems necessary. At this point of our investigation, it is obvious that, on one hand, implementing ferrocene breaks the electronic coupling due to a relatively poor π-electron density. Important is the comparison with porphyrins or exTTF. On the other hand, the choice of ethynylene as a connection point to the Fc donor is malapropos in the sense, that the shorter bond length and the polarizing character of the triple bond hinder the system from establishing a homogeneously extended π-conjugation.

Nonetheless, based on the obtained results and the variety of the investigated structures the importance of fully extended π-conjugation has been shown as the crucial factor for efficient energy- and charge-transfer properties. Especially, in view of distance-independent electron-transport phenomena the homogeneous distribution of electron density and the matching of the energy levels of the single modular components is of utmost importance. In this context, our data shows, that oPPVs exhibit the best charge-transfer characteristics, followed by oFLs and oPPEs. In spite of the fact that the triple bonds in the oPPE oligomers were supposed to increase the rigidity of the systems and, in turn, improve the π-conjugation by forcing the configuration into planarity due to higher rotational barriers, they seem to interrupt the π-system due to their shorter bond lengths and polarizing character.

In the end, still many more questions remain to be answered, the most important of which is how to integrate the molecular wires into electronic and optoelectronic devices of the future. Such integration efforts raise many problems, which likely will require several iterations in molecular wire research. For instance, challenges that are related to the controlability of the rate processes by synthetic efforts, positioning and excitation. Certainly, these challenges will continue making molecular wires posing tantalizing tasks to the science and technology communities in the future.

As it was pointed out, the beauty of organic chemistry is that simple changes used during the synthesis of molecular wires yields products with vastly different physical properties. Even minor alterations, such as the exchange of the donor moiety turned out to strongly affect the conduction behavior of the wires. Hence, the first molecular wire(s) that eventually might appear in commercial devices may bear no resemblance with those we have discussed. Thus, research in this area can still yield much fruit.

Curriculum Vitae

Personal

Name: Mateusz Barnaba Wielopolski

Place of Birth: Katowice (Poland)

Date of Birth: June 12th, 1981

Family Status: Single, One child

Nationality: German

Tertiary Education

June, 2006 – December, 2008: Dissertation on "TestingMolecular Wires - A Photophysical and Quantum Chemical Assay" at the Department of Physical Chemistry and the Computer Chemie Centrum in the groups of Prof. Dr. Dirk M. Guldi and Prof. Dr. Timothy Clark at the Friedrich-Alexander-University in Erlangen

October, 2004 – May, 2006: Academic Studies of Molecular Nano Science at the Friedrich-Alexander-University in Erlangen, Degree: Master of Science, Grade: 1.1Master Thesis: "Molecular Wires - Relationship between Chemical Nature and Long-Range Electron Transfer"

October, 2001 – September, 2004: Academic Studies of Molecular Science at the Friedrich-Alexander-University in Erlangen, Degree: Bachelor of Science Grade: 1.8, Bachelor Thesis: "Investigations on the Mechanism of the Extrusion of Atomic Nitrogen from the Interior of a C_{60} Cage"

Scholar Education

September, 1991 - June, 2000:   Highschool leaving examination at the Theresiengymnasium in Ansbach (Germany) Major courses: Chemistry, English, Grade: 1.4

August, 1997 - December, 1997:   Highschool in Lethbridge, Alberta (Canada) as a 3-month student exchange

Publications

S. Shankara Gayathri, Mateusz Wielopolski, Emilio M. Pérez, Gustavo Fernández, Luis Sánchez, Rafael Viruela, Enrique Ortí, Dirk M. Guldi, Nazario Martín. Discrete Supramolecular Donor-Acceptor Complexes, *Angew. Chem. Int. Ed.* **2008**, *48*, 815.

Renata Marczak, Mateusz Wielopolski, S. Shankara Gayathri, Dirk M. Guldi, Yutaka Matsuo, Keiko Matsuo, Kazukuni Tahara, Eiichi Nakamura. Uniquely Shaped Double-Decker Buckyferrocenes-Distinct Electron Donor-Acceptor Interactions. *J. Am. Chem. Soc.* **2008**, *130*, 16207.

Jose Santos, Beatriz M. Illescas, Mateusz Wielopolski, Ana M. G. Silva, Augusto C. Tome, Dirk M. Guldi, Nazario Martín. Efficient Electron Transfer in β-substituted Porphyrin-C_{60} Dyads Connected Through a p-Phenylenevinylene Dimer. Tetrahedron **2008**, *64*, 11404.

Mateusz Wielopolski, Carmen Atienza, Timothy Clark, Dirk M. Guldi, Nazario Martín. p-Phenyleneethynylene Molecular Wires: Influence of Structure on Photoinduced Electron-Transfer Properties. *Chem. Eur. J.* **2008**, *14*, 6379.

Jan-Frederik Gnichwitz, Mateusz Wielopolski, Kristine Hartnagel, Uwe Hartnagel, Dirk M. Guldi, Andreas Hirsch. Cooperativity and Tunable Excited State Deactivation: Modular Self-Assembly of Depsipeptide Dendrons on a Hamilton Receptor Modified Porphyrin Platform. *J. Am. Chem. Soc.* **2008**, *130*, 8491.

Carmen Atienza-Castellanos, Mateusz Wielopolski, Dirk M. Guldi, Cornelia van der Pol, Martin R. Bryce, Salvatore Filippone, Nazario Martín. Determination of the Attenuation Factor in Fluorene-Based Molecular Wires. *Chem. Commun.* **2007**, *48*, 5164.

Cornelia van der Pol, Martin R. Bryce, Mateusz Wielopolski, Carmen Atienza-Castellanos, Dirk M. Guldi, Salvatore Filippone, Nazario Martín. Energy Transfer in Oligofluorene-C_{60} and C_{60}-Oligofluorene-C_{60} Donor-Acceptor Conjugates. *J. Org. Chem.* **2007**, *72*, 6662.

Carmen Atienza, Nazario Martín, Mateusz Wielopolski, Naomi Haworth, Timothy Clark, Dirk M. Guldi. Tuning Electron Transfer Through p-Phenyleneethynylene Molecular Wires. *Chem. Commun.* **2006**, *30*, 3202

Conference Contributions

Talks

Molecular Wires—Electron Transfer in π-Conjugated Donor-Acceptor Systems. *22nd Darmstädter Molecular Modelling Workshop Erlangen, Germany*, April **2008**.

C_{60}-WIRE-exTTF—Molecular Wires with Small Attenuation Factors. *3rd European Young Investigator Conference, Slubice, Poland*, June **2007**.

C_{60}-WIRE-exTTF—Molecular Wires with Small Attenuation Factors. *EUROCORES-SONS Workshop—Molecular Nanoelectronics, Veilbronn, Germany*, March **2007**.

Molecular Wires with Small Attenuation Factors Based on C_{60}-WIRE-exTTF. *Rennes-Erlangen Symposium of Chemistry, Rennes, France*, June **2006**.

Poster

Molecular Wire Behavior of Organic π-Conjugated Oligomers in Donor-Wire-Acceptor Conjugates. In Proceedings of the 22nd IUPAC Symposium on Photochemistry. *Gothenburg, Sweden*, July **2008**.

Molecular Wire Behavior of Organic π-Conjugated Systems in Donor-Wire-Acceptor Conjugates. In Proceedings of the 22nd Darmstädter Molecular Modelling Workshop. *Erlangen, Germany*, April **2008**.

Molecular Wire Behavior of Organic π-Conjugated Systems in Donor-Wire-C_{60} Conjugates. In Proceedings of the Central European Conference on Photochemistry. *Bad Hofgastein, Austria*, February **2008**.

Electron-Transfer Systems Based on C_{60}-Theory and Experiment. In Proceedings of the 21st Darmstädter Molecular Modelling Workshop. *Erlangen, Germany*, May **2007**.

Molecular Wires Based on C_{60}-WIRE-exTTF. In Proceedings of the 20th Darmstädter Molecular Modelling Workshop. *Erlangen, Germany*, May **2006**.

Molecular Wires Based on C_{60}-WIRE-exTTF. In Proceedings of the 20th Darmstädter Molecular Modelling Workshop. *Erlangen, Germany*, May **2006**.

Molecular Wires Based on C_{60}-WIRE-exTTF. In Proceedings of the 105th General Meeting of the Bunsen-Community for Physical Chemistry. *Erlangen, Germany*, May **2006**.